ENGELS AS MILITARY CRITIC

ENGELS AS MILITARY CRITIC

Articles by Friedrich Engels
reprinted from the *Volunteer Journal*
and the *Manchester Guardian* of the 1860s

with an introduction by

W. H. CHALONER

AND

W. O. HENDERSON

GREENWOOD PRESS, PUBLISHERS
WESTPORT, CONNECTICUT

Library of Congress Cataloging in Publication Data

Engels, Friedrich, 1820-1895.
Engels as military critic.

Reprint of the ed. published by Manchester University Press, Manchester, Eng.
Includes index.
1. Military art and science--Collected works.
2. Military history--Collected works. I. Title.
U17.E5 1976 355'.008 75-26213
ISBN 0-8371-8407-X

This edition published in 1959 by Manchester University Press, Manchester

Reprinted with the permission of Manchester University Press

Reprinted in 1976 by Greenwood Press, a division of Williamhouse-Regency Inc.

Library of Congress Catalog Card Number 75-26213

ISBN 0-8371-8407-X

Printed in the United States of America

CONTENTS

PART IV
THE CIVIL WAR IN THE UNITED STATES OF AMERICA

PART V
THE SCHLESWIG-HOLSTEIN WAR, 1864

PART VI
THE SEVEN WEEKS' WAR, 1866

ACKNOWLEDGEMENTS

THE editors of this selection of Friedrich Engels's writings on military affairs desire to thank Mr. V. I. Tomlinson for the loan of a complete set of the *Volunteer Journal for Lancashire and Cheshire*; Mr. S. B. Haw for the loan of M. Bennett's *The Story of the Rifle*; the editor of the *Manchester Guardian* for permission to reprint Engels's letter on the Schleswig-Holstein War and his articles on the Seven Weeks' War which appeared in that newspaper; Mr. T. L. Jones, the Secretary of the Manchester University Press, for much assistance in seeing the book through the press; the secretaries of the History Department of the University of Manchester for preparing the typescript; and our colleague Mr. A. E. Musson for his help in reading the proofs.

W. H. C.
W. O. H.

NOTE

Engels's articles have been reprinted in a different order from that in which they originally appeared in the *Volunteer Journal for Lancashire and Cheshire.* This has been done in order to bring together the articles on the following topics—the Volunteers, the rifle, the French Army, and the American Civil War. Information concerning persons mentioned by Engels will be found in the index. The spelling of some proper names has been modernized. Editors' footnotes to Engels's text are enclosed in square brackets.

INTRODUCTION

The significance of the part played by Friedrich Engels as a pioneer of 'scientific' Socialism is well known. As a young man he contributed to the *Rheinische Zeitung*, the Cologne radical paper which Karl Marx edited in the early 1840s. Engels's visit to Manchester in 1842–4 enabled him to collect material for his book on *The Condition of the Working Class in England*[1] which established his position in Socialist circles as an authority on the social consequences of industrialization. His early activities as a revolutionary Socialist began with propaganda among the workers in his native city of Barmen (Rhineland) in the winter of 1844–5 and ended when he fled to Switzerland after the collapse of the abortive rising in Baden in 1849. By that time he was already closely associated with Karl Marx and the two young men had co-operated in drawing up the Communist Manifesto (1847–8) and in editing the *Neue Rheinische Zeitung* in Cologne (1848–9).[2]

By the end of 1850 Engels was back in Manchester, where he worked until 1870 with the cotton firm of Ermen and Engels.[3] He eventually became a partner in the firm. The work was uncongenial but he stuck to his post so as to earn enough money to contribute towards the support of Marx, who was working in London on the first volume of *Das Kapital*. During those twenty years he maintained a regular correspondence with Marx and the two men discussed many aspects of Socialist doctrine. After 1870 Marx and Engels both lived in London and the close collaboration between them continued. When Marx died in 1883 many Socialists regarded Engels as the 'father' of their movement.

Engels was a man of many parts. He earned his living for twenty years as a businessman in Manchester. He wrote articles regularly for the *New York Daily Tribune* and other papers on a great variety of topics. He kept up a correspondence with Marx

[1] Friedrich Engels, *The Condition of the Working Class in England* (translated by W. O. Henderson and W. H. Chaloner, 1958).

[2] See Karl Marx (ed.), *Neue Rheinische Zeitung* (reprint of 1955).

[3] For Engels's twenty years' residence in Manchester see W. O. Henderson and W. H. Chaloner, 'Friedrich Engels in Manchester', (*Memoirs and Proceedings of the Manchester Literary and Philosophical Society*, Vol. 98, 1956–7).

in which numerous problems connected with the writing of *Das Kapital* were discussed at length. He kept in touch with many revolutionary Socialists who were living in exile in England and America. He was a linguist of no mean order.

One subject in which Engels was particularly interested was the art of warfare. His writings on military affairs eventually earned him the nickname of 'the General' from his political associates.[1] In his youth he had served as a volunteer in the Guards Artillery of the Prussian army for a year in Berlin (October, 1841—September 30, 1842). His letters to his sister suggest that Engels did not take his military duties very seriously. He was not above malingering and once succeeded in avoiding a route march at night by pretending to have toothache.[2] He succeeded, however, in securing promotion to the rank of *Bombardier*.[3]

In the obituary notice of Engels published in the *Newcastle Weekly Chronicle* of August 17, 1895, George Julian Harney wrote '. . . that experience [of military service] did not increase his admiration of the Prussian military system or the Prussian Government.'

In 1849 Engels took part in a minor insurrection in the Bavarian Palatinate and in Baden.[4] He gave a brief description of his part in the affair in a letter to Frau Marx written from Switzerland on July 25, 1849.[5] He explained that he joined the rebels at

[1] Gerhard Zirke, *Der General: Friedrich Engels, der erste Militär-theoretiker der Arbeiterklasse* (Gesellschaft zur Verbreitung Wissenschaftlicher Kenntnisse, Leipzig and Jena, 1957).

[2] F. Engels to his sister Marie, August 2–8, 1842, in *Marx-Engels Gesamtausgabe*, Part I, Vol. 2 (1930), p. 623.

[3] F. Engels to his sister, April 14–16, 1842, *ibid.*, p. 620. Engels was eventually drummed out of the Prussian *Landwehr* as a deserter in 1860 (Engels to Marx, January 31, 1860, in *Marx–Engels Gesamtausgabe*, Part III, Vol. 2 (1930), p. 459).

[4] For a brief account of this rising see Gustav Mayer, *Freidrich Engels* (abridged English translation, 1936), pp. 110–13. A list of books and articles on the insurrection appears in Gustav Mayer, *Friedrich Engels* (revised German edition, 2 vols., 1934), Vol. I, p. 393.

[5] Friedrich Engels to Frau Jenny Marx, July 25, 1849, in *Marx–Engels Gesamtausgabe*, Part III (*Briefwechsel*) (Correspondence), Vol. I (1929), p. 109. Cited in future as *Gesamtausgabe*. See also F. Engels, 'Die deutsche Reichsverfassungskampagne' in Karl Marx (ed.), *Neue Rheinische Zeitung* (reprint of 1955), pp. 94–104, 149–74, and Klaus Schreiner, *Die badisch-pfälzische Revolutionsarmee, 1849* (Berlin, 1956).

Kaiserslautern and acted as adjutant to August Willich[1] from June 13 to July 12 when the Prussian forces drove the insurgents over the frontier into Switzerland. Engels took part in four engagements of which only one—at Rastatt—was of any importance. He wrote four articles on the campaign in South Germany for Marx's *Neue Rheinische Zeitung* of 1850.

Three years later, in an article published in the *New York Daily Tribune*, Engels discussed the rising in which he had taken part. He criticized the conduct of the initial operations by Franz Sigel, who was in command at the beginning of the campaign.[2] Engels wrote:

'As to the fighting part of the business, never were military operations carried on in a more slovenly, more stolid way than under the Baden General-in-Chief Sigel, an ex-lieutenant of the regular army. Everything was got into confusion, every good opportunity was lost, every precious moment was loitered away with planning colossal, but impracticable projects, until, when at last the talented Pole Mieroslawski,[3] took up the command, the army was disorganized, beaten, dispirited, badly provided for, opposed to an enemy four times more numerous, and withal, he could do nothing more than fight at Waghäusel, a glorious but unsuccessful battle, carry out a clever retreat, after a last hopeless fight under the walls of Rastatt, and resign.'[4]

Despite the limited nature of his experience under fire Engels soon began a serious study of military history and strategy. In the spring of 1851, when he was back at his office stool in Manchester, he told Marx that he was finding difficulties in borrowing General Sir William Napier's *History of the War in the Peninsula* from the library of the Athenaeum.[5] The project of a study of the military

[1] August Willich (1810–78)—a former lieutenant in the Prussian Army—later emigrated to the United States and fought on the side of the North during the Civil War.

[2] Franz Sigel (1824–1902) subsequently emigrated to the United States, where he became General in command of the Army of West Virginia in the Civil War.

[3] L. Mieroslawski (1814–78) was in command from June 10 to July 3, 1849, when Sigel was again appointed to take charge of the troops.

[4] Karl Marx (should be F. Engels), *Revolution and Counter Revolution or Germany in 1848* (edited by Mrs. Eleanor Marx Aveling, 1937), p. 128.

[5] F. Engels to K. Marx, March 17, 1851, in *Gesamtausgabe*, Part III, Vol. I, p. 169.

campaign associated with the revolutions of 1848–9 was discussed by the two friends but was not carried out.[1] Instead Engels wrote a series of nineteen or twenty articles of a more general character on the revolutionary movement in Germany and the subsequent reaction. These articles were written in English by Engels, who sent them to Marx. They were then forwarded to C. A. Dana of the *New York Daily Tribune* for publication in that paper under Marx's name. Dana supposed that they were written by Marx and as late as 1891—when the articles were reprinted as a book—Mrs. Aveling (Marx's daughter) thought that they had been written by her father. It is clear, however, from the Marx–Engels correspondence that Engels alone was the author.

Marx and Engels became regular contributors to the *New York Daily Tribune* and wrote on a variety of topics of current interest. The paper, which had a circulation of 200,000, was one of the most influential in the United States. By the autumn of 1852 Marx was writing articles in German which Engels translated into English. Early in 1853, however, Marx felt sufficiently confident of his command of the language to write an article for Dana in English.[2] But he still sometimes asked Engels to translate into English articles which he had written in German. Engels usually wrote on military topics while Marx discussed economic, political and industrial problems. On the eve of the Crimean War Marx thanked Engels for a 'beautiful article' which, he said, would secure for Dana the reputation of 'a Field Marshal'.[3]

Since Marx and Engels had been writing for the *New York Daily Tribune* for some years it is not surprising that when C. A. Dana and George Ripley decided in 1857 to bring out a *New American Cyclopaedia* they should have invited Karl Marx to become one of the contributors.[4] Marx asked Engels to write some of the articles on military subjects. The Crimean War was

[1] K. Marx to F. Engels, April 2, 1851, and F. Engels to K. Marx, April 3, 1851, in *Gesamtausgabe*, Part III, Vol. I, pp. 179–82.

[2] K. Marx to F. Engels, January 29, 1853, in *Gesamtausgabe*, Part III, Vol. I, pp. 444–5.

[3] K. Marx to F. Engels, December 2, 1853, in *Gesamtausgabe*, Part III, Vol. I, p. 514.

[4] See *Gesamtausgabe*, Part III, Vol. 2, pp. 196–7, 200–1, 217–18, 221–4, 231, 237, 267, 270, 276, 228, 282–3, 290, 300–1, 336, 418, 419, 421.

over[1] and there were no campaigns in progress about which Engels could write. In April 1857, he told Marx that he found peace demoralizing and that he looked forward to spending his evenings writing articles on military topics for the new encyclopaedia. Moreover Dana's offer of a fee of two dollars a page was welcomed by Marx who was, as usual, desperately short of money.[2] The articles written by Engels included some general surveys,[3] some military biographies,[4] and some descriptions of battles.[5]

The international tension in 1859 which came to a head in the Italian war caused Engels to turn his attention again to current military problems—particularly to the question of the extent to which Germany's security was affected by the threat of hostilities between France and Austria. He had written hardly anything for the German public since the days when he had contributed to the *Neue Rheinische Zeitung* during the revolution of 1848–9.[6] The reactionary forces appeared to be loosening their grip upon his native country. Frederick William IV's health broke down in the autumn of 1857 and his brother Prince Wilhelm became Regent in October, 1858. A short-lived 'New Era' in Prussian politics began with the dismissal of Otto von Manteuffel as Minister President. Engels hoped to gain the ear of the German public once more. And this he succeeded in doing. His pamphlet on *Po und Rhein* was published anonymously in Berlin in 1859 but the name of the author was soon revealed by Marx in his paper *Das Volk* (published in London). In a letter of January 11, 1860, Marx congratulated Engels on the appearance of a very favourable review of his pamphlet in a leading German military journal (the *Allgemeine Militärzeitung*, published at Darmstadt). He added: 'Your pamphlet has established your position in

[1] For the views of Marx and Engels on the Crimean War see Karl Marx, *The Eastern Question: a reprint of letters written 1853–1856* (edited by Eleanor Aveling and Edward Aveling, 1897) and Rjasanow (ed.), *Gesammelten Schriften von Marx und Engels, 1852–62* (2 vols., [covering 1852–55] 1917).

[2] F. Engels to K. Marx, April 22, 1857, in *Gesamtausgabe*, Part III, Vol. 2, pp. 185–7.

[3] E.g. articles on 'Army', 'Cannon', 'Fortification', 'Infantry' and 'Navy'.

[4] E.g. article on 'Blücher'.

[5] E.g. articles on 'the Armada' and 'Aboukir'.

[6] Two articles by Engels on 'Germany and Panslavism' had appeared in the Breslau newspaper *Neue Oder Zeitung* on April 21 and 24, 1855.

Germany as a military critic.'[1] A sequel entitled *Savoyen, Nizza und der Rhein* appeared in Berlin in 1860.[2]

Events on the Continent in 1858–9 caused alarm in England and it was widely believed that Napoleon III was planning to invade this country. On May 12, 1859, General Peel, the Secretary of State for War, circularized the Lord Lieutenants and informed them that local Volunteer Corps might be formed under an Act of 1804 (44 Geo. III cap. 54). The plan received enthusiastic public support and by the end of 1860 some 119,000 volunteers had been enrolled.[3] The volunteers were a very heterogenous 'army'. In the rural districts and market towns the gentry vied with each other in raising corps largely composed of their tenants. In the manufacturing districts industrialists and merchants formed corps which were joined by their employees. In Liverpool in 1860 there existed a 'Borough Guard' (led by Joseph Mayer, the antiquarian, and composed mainly of tradesmen), an 'Irish Corps', a kilted 'Scots Corps', a 'Press Guard' of journalists who wore red shirts, an 'Artisans Corps', and the 'Artillery Volunteers' (under James Walter).[4] At Nottingham the Liberal industrialist A. J. Mundella took the initiative in forming a local corps in May, 1859, while in Bradford, W. E. Forster, the Liberal politician, himself drilled a company raised from among his own mill-workers.[5]

In addition to local corps raised on a territorial basis the Engineer and Railway Staff Corps was set up in 1865 to mobilize railway transport in wartime. The well-known engineer Charles Manby played an important part in setting up this corps which was commanded by W. M. McMurdo.[6] That the problem of the part to be played by the railways in mobilization had been

[1] K. Marx to F. Engels, January 11, 1860 in *Gesamtausgabe*, Part III, Vol. 2, p. 453. Engels had originally intended to give this pamphlet the subtitle 'Military Studies' but, on Marx's advice, he did not do so.

[2] The two pamphlets (edited by E. Bernstein) were reprinted by Dietz at Stuttgart in 1915. For a discussion of the pamphlets see G. Mayer, *Friedrich Engels* (German edition, two vols., 1934), Vol. II, ch. 3.

[3] See Sir W. M. McMurdo's article on 'Volunteers' in the *Encyclopaedia Britannica* (9th edition), Vol. XXIV (1888), pp. 294–6.

[4] R. B. Rose, 'The Liverpool Volunteers of 1859' in the *Liverpool Bulletin*, Vol. VI, Numbers 1 and 2, October 1956, pp. 47–66.

[5] W. H. G. Armytage, *A. J. Mundella, 1825–1897: the Liberal Background to the Labour Movement* (1951), pp. 29–30.

[6] E. A. Pratt, *The Rise of Rail-Power in War and Conquest, 1833–1914* (1915), pp. 179–82.

discussed for some time is clear from Engels's remarks in his article on 'Volunteer Engineers'.[1]

On November 26, 1859, Marx told Engels that, in an article written for the *New York Daily Tribune*, he had promised to give an account of the English Volunteer Movement on some future occasion. He asked Engels to supply him with the necessary information.[2] An article by Engels on a review of the Volunteers held on the Newton-le-Willows racecourse in Lancashire on August 11, 1860, appeared in the *Allgemeine Militärzeitung* (Darmstadt) in its issue of September 8. This article, translated into English, was reprinted in the *Volunteer Journal for Lancashire and Cheshire* on September 14. This local weekly journal had only recently been established in Manchester. It was edited by Captain Isaac Hall and published by W. H. Smith & Sons, who sold it on their railway bookstalls.

Marx wrote to Engels on October 2, 1860: 'Your rifle article had made the rounds of the entire London press and had also been reviewed in the *Observer*, which reflects the views of the Government. It was sensational.'[3] Engels complacently replied that this had not been achieved without some effort on his part. He had sent copies of the *Volunteer Journal for Lancashire and Cheshire* to all the London newspapers with his own article marked with red pencil. In a covering note to the various editors he had pointed out that 'as this is the first professional opinion of a foreign military paper on the voluntary (*sic*) movement it may be of interest'.[4]

This was the first of twenty-nine articles contributed by Engels to the *Volunteer Journal* between September, 1860, and March, 1862. Fourteen of them—including eight on the history of the rifle[5]—were reprinted in Manchester in 1861 by W. H.

[1] See below, pp. 23–4.

[2] K. Marx to F. Engels, November 26, 1859, in *Gesamtausgabe*, Part III, Vol. 2, p. 442.

[3] K. Marx to F. Engels, October 2, 1860, in *Gesamtausgabe*, Part III, Vol. 2, p. 514.

[4] F. Engels to K. Marx, October 5, 1860, in *Gesamtausgabe*, Part III, Vol. 2, p. 515.

[5] For a recent account of the early history of the rifle see T. H. McGuffie, 'Musket and Rifle', Parts 1 and 2 (*History Today*, Vol. VII, 1957, pp. 257–63 and 473–9). See also M. Bennett, *The Story of the Rifle* (Cobbett Publishing House, n.d. (1944)).

Smith as a sixpenny pamphlet entitled *Essays addressed to Volunteers*. In a preface to this pamphlet, dated March 9, 1861,[1] Engels observed that although the facts which he assembled were 'neither novel nor original', his conclusions and opinions were his own. A review of the pamphlet appeared in the *United Services Gazette* of March 23, 1861. The reviewer observed:

'We read the "History of the Rifle" with much pleasure, and certify to its accuracy in all important particulars. "The French Light Infantry" we did not like quite so well, as the tone indicates that the writer is, to a considerable extent, bitten with that new-fangled admiration for French soldiering which we, after long and intimate knowledge, hold to be an utter delusion. The paper, however, is valuable . . . The question of Volunteer Artillery is well handled, and . . . the whole *brochure* . . . is modestly and carefully written.'

Engels's articles on the rifle appear to have been written with a view to publication in the *New York Daily Tribune*[2] but they eventually appeared in the *Volunteer Journal*. The articles which Engels wrote for the *Volunteer Journal* dealt not only with the Volunteer movement in England and with the history of the rifle but also with other military topics such as the French army and the American Civil War. The present edition includes a full reprint of all the articles contributed by Engels to the *Volunteer Journal*.[3]

[1] The preface was signed with the initials 'F.E.' (There is a brief reference to the pamphlet in a letter written by Engels to Marx on February 6, 1861, in the *Gesamtausgabe*, Part III, Vol. 3, p. 12.) In the *Volunteer Journal* Engels's article can be identified in various ways. The first was stated to be translated from the Darmstadt *Allgemeine Militärzeitung* and the author of that article can be identified from the statement made by Engels in his letter to Marx of October 5, 1860. Subsequent articles can be identified in various ways. Some were initialled, others were attributed to the author of previous articles or of the pamphlet (known to have been written by Engels).

[2] On August 1, 1860 Engels wrote to Marx: 'You might ask the [*New York Daily*] *Tribune*, if they would like to have a series of four or five articles on rifled firearms including all the latest improvements. I'm not writing them on speculation' (*Gesamtausgabe*, Part III, Vol. 2, p. 500).

[3] Engels also translated Marshal Bugeaud's views on the principles of fighting, which appeared as three articles in the *Volunteer Journal* for February 9, and 16, and March 2, 1861 under the title 'On the moral element in fighting' (see p. 103 below). He also contributed an article on Bugeaud to the *New American Cyclopaedia*. Engels's *Volunteer Journal* articles have been

In the early 1860s Engels co-operated with Marx in writing on the American Civil War for the *New York Daily Tribune* and the Vienna *Presse*. Engels was unduly optimistic concerning the military prospects of the Southern forces, but Marx saw more clearly the immense resources of strength on which the North could ultimately rely. These articles by Marx and Engels, together with the extracts from their correspondence, were reprinted in 1938 under the title *The Civil War in the United States* (edited by R. Enmale). During the Schleswig-Holstein war Engels wrote a letter to the *Manchester Guardian* (February 16, 1864)—under the initials 'F. E.'—on the size of the forces in Schleswig and this letter has been reprinted in the present edition of Engels's writings on military subjects. In 1865 he wrote a pamphlet entitled *Die Preussische Militärfrage und die Deutsche Arbeiterpartei* which offered advice to the German workers on the attitude that they should adopt towards the expansion and reorganization of the Prussian army.

In the summer of 1866 Engels contributed five articles to the *Manchester Guardian* on the Seven Weeks' War.[1] Like other military experts he greatly underrated both the efficiency of the Prussian army and the skill with which Moltke handled a situation which that commander himself described as 'unfavourable but inevitable'.[2] Engels was surprised at the rapid and overwhelming success of the Prussians at Königgrätz. These 'Notes on the War' are reprinted below.

During the Franco-Prussian War Engels wrote sixty articles for the *Pall Mall Gazette* on the campaigns of 1870–1. These articles were reprinted in Vienna in 1923 under the title *Notes on the War. Sixty articles reprinted from the 'Pall Mall Gazette'* (edited by F. Adler).

Engels's standing as a military critic may be judged from the fact that several leading newspapers in England (*Pall Mall Gazette*, *Manchester Guardian*) and abroad (*New York Daily*

reprinted in Russian in the collected edition of Marx and Engels's writings published by the Marx-Engels-Lenin Institute (*Sotchinenia*, vol. XII, part 2 (Moscow, 1934), pp. 403–531).

[1] The articles appeared anonymously. They are attributed to Engels by his biographer Gustav Mayer in *Friedrich Engels* (German edition, 2 vols., 1934), Vol. II, pp. 150–4. See also Maximilien Rubel, *Bibliographie des œuvres de Karl Marx* . . . (1956), p. 248.

[2] Heinrich Friedjung, *The Struggle for Supremacy in Germany, 1859–66* (abridged English translation, 1935), p. 213.

Tribune, *Allgemeine Militärzeitung*) printed his articles. As a writer on military subjects Engels made a name for himself by his considerable knowledge of the history of war—evidence for this is to be found in the articles which he wrote for the *New American Cyclopaedia*—and also by the infinite pains which he took to secure accurate information concerning the military problems of his day. When he began his studies on military affairs he told Marx that 'nothing is easier than to make a blunder when writing critical accounts of campaigns, if one has not had access to all the data concerning the number of troops involved, how they are provisioned, and what munitions they have'.[1] When he erred—his mistakes include a failure to appreciate the strength of the North in the American Civil War and of the Prussians in the Seven Weeks' War—he erred in company with other experts on military affairs.

Engels devoted his life to the cause of revolutionary Socialism. It may be asked whether the time that he devoted to the study of the art of warfare was merely an agreeable recreation or whether he felt that it was his duty as a good Socialist to keep abreast with the military thinking of his day. The climax to Engels's early career as a revolutionary agitator came in 1848–9 when he worked with Marx on the *Neue Rheinische Zeitung* and took part in the Baden rising. The triumph of reaction was a bitter blow, and in exile in England he and Marx brooded over the failure of the revolutionaries in 1848–9. Engels's plan to study the Magyar and other risings had a practical purpose. By discovering why the risings had failed it might be possible to save future revolutionary risings from a similar fate. Engels's interest in the campaigns of 1848–9 broadened into a wider study of military history.

He was convinced that revolution was bound to come. Even in 1845, when writing his book on *The Condition of the Working Class in England*, he had expressed the view that capitalism in England was doomed. The next slump—he foresaw that it would come in 1847—would, he believed, herald the collapse of the social system. In fact the Chartist demonstration in 1848 was a complete failure. Engels was not discouraged. Whenever there was a depression in trade—as in 1857—he continued to hope for an outbreak of violence on the part of the workers which

[1] F. Engels to K. Marx, April 3, 1851, in *Gesamtausgabe*, Part III, Vol. 2, p. 182.

would sweep away the moneyed classes. His fundamental interest in the Volunteer movement may well have sprung from the hope that one day the English bourgeoisie and not the French *chasseurs* would have to face the rifles of the Volunteers.

Marx welcomed Engels's interest in military affairs for another reason. In a letter of September 25, 1857, he wrote:

'Nowhere is the relationship between factors of production and the structure of society more clearly illustrated than in the history of the army. Economic expansion is greatly influenced by the army. In ancient times, for example, the payment of wages in money was first fully developed in the army. In Roman law the *peculium castrense* was the first legal recognition of the principle that property could be owned by any one other than the head of a family. The same applies to the gild activities of the corporation of blacksmiths [in Roman times]. And after what Grimm calls the Stone Age the special value given to metals and their use as money appears to rest upon their importance in time of war. Moreover the division of labour within a single sphere of human activity was first seen in military forces. The whole history of the structure of middle-class society is clearly summarized in the history of armies.' [1]

The following selection of Friedrich Engels's articles on military affairs represents only a part of his studies on the art of war.[2] It includes all his contributions to the *Volunteer Journal* and also what he wrote for the *Manchester Guardian* on Prussia's wars against Denmark (1864) and Austria (1866). While Engels's writings on the Crimean War, the American Civil War, and the Franco-Prussian War have already been reprinted his contributions to the *Volunteer Journal* have fallen into undeserved oblivion. His articles are of particular interest in the centenary year of the founding of the Volunteer movement. They give the impressions of a well-informed German critic on an important episode in British military history.

W. H. Chaloner
W. O. Henderson

[1] K. Marx to F. Engels, September 25, 1857, in *Gesamtausgabe*, Part III, Vol. 2, p. 228. This point of view was later elaborated by W. Sombart in his well-known book on *Krieg und Kapitalismus* (1913).

[2] The East German Ministry of National Defence has published: F. Engels, *Ausgewählte militärische und militär-politische Schriften*, 2 vols. (1957).

PART I
THE VOLUNTEER MOVEMENT

1. A Review Of English Volunteer Riflemen[1]

The *Allgemeine Militärzeitung*, published at Darmstadt, and considered the first military paper in Germany, in its number of the 8th September, gives an account by a correspondent of the Newton Review and of the rifle movement in general. The following is a translation of this article (prepared specially for the *Volunteer Journal*) which no doubt will prove interesting to the volunteers of Lancashire and Cheshire and especially to those who were present at the review. As may be expected, this account is not made up of that unqualified praise which the British press generally gives as its contribution to the movement; still the character of the contemporary in question ought to be a sufficient guarantee that it is not written by an incompetent hand, and the sympathetic tone of the whole article proves that the writer had no inclination for wanton fault-finding. As to the suggestions contained in the article, we shall leave our readers to form their own opinion upon them.' [2]

England, as well as Germany, is arming to repel the attack with which Bonapartism threatens her; the British Volunteer Riflemen arose for the same cause which made Prussia double the number of her battalions of the line. It will, therefore, be of interest to the German military public to receive some detailed information on the present state and the fitness for actual service of the British Volunteer Army; for this army, from its very origin, and in virtue of its fundamental idea, is an enemy of Bonapartism, an ally of

[1 *Volunteer Journal for Lancashire and Cheshire*, September 14, 1860, Vol. I, No. 2, pp. 26–8. Translated from the *Allgemeine Militärzeitung*, Darmstadt September 8, 1860. The original title of this article was—'A German account of the Newton Review' but Engels altered it to 'A Review of English Volunteer Riflemen' in his *Essays addressed to Volunteers* (1861).]

[2 This introductory paragraph was written by the Editor of the *Volunteer Journal.*]

Germany. A very few battalions excepted, this army of volunteers dates from the latter half of last year; the great body has not been put in uniform and drilled more than a twelvemonth. At present, its strength, on paper, is 120,000 men; but if we may draw conclusions from what is the fact in some districts, there will not be more than 80,000 men really effective and drilled; the remainder take no interest in the matter, and had better be erased from the lists.

The organization is very simple. Wherever 60 to 100 volunteers (in the artillery 50 to 80) are brought together, in any locality, they form themselves into a company, subject to the consent of the Lord Lieutenant of the county. They elect candidates for officers (a captain, a lieutenant, and an ensign), whom the Lord Lieutenant, in most cases, appoints to their respective companies; but there have also been instances of rejection. Several companies may form themselves into a battalion, in which case the Lord Lieutenant appoints the major and lieutenant-colonel, mostly according to the wishes of the officers, or according to the seniority among the captains. Thus there are corps varying from one to eight companies and more, numbered in the order of their formation in their respective counties; but only full battalions of eight companies receive a lieutenant-colonel. The officers may, all of them, be appointed from among the volunteers, and they are not subjected to any examination. The adjutant, however, must be an officer from the line of militia, and he alone receives regular pay. The volunteers find their own clothing, etc., but if desired, the government furnishes them with rifle and bayonet by way of loan. The colour and cut of the uniform is fixed by the various corps, subject to the approval of the Lord Lieutenant. The corps have also, upon the whole, to find their own drill and practice grounds, ammunition, instructors and music.

The uniforms of the various infantry or rifle corps are mostly dark green, dark or light grey, or brown drab. The shape is something intermediate between the French and English pattern; for a headdress they mostly wear the French *képi*, or the French or English officer's cap. The artillery is dressed in dark blue, and has adopted, for appearance's sake, the rather unserviceable and lumbering fur-cap or busby of the horse artillery. There are also a few mounted rifles whose uniform imitates that of the English cavalry, but they are a mere article of luxury.

At the time when the formation of these rifle corps was first agitated, the whole matter savoured very strongly of our own national and civic guards; there was a great deal of playing at soldiers; the way in which officers were manufactured, and the appearance and helplessness of some of these officers, when on duty, were rather amusing. It may well be imagined, the men did not always elect the most capable, or even those who had the movement most at heart. During the first six months, almost all battalions and companies made the same effect upon the beholder as our own defunct civic guard of 1848.

This, then, was the material handed over to the drill-sergeants, in order to shape it into a body of serviceable field-troops. The manual and platoon was gone through mostly at nights, between seven and nine o'clock, in covered rooms and by gas-light, twice or three times a week. On Saturday afternoons, if possible, the whole body made a short march and went through company movements. To drill on Sunday was forbidden by both law and custom. The instructors were sergeants and corporals of the line, the militia, or pensioners; and they, too, had to form the officers into shape. But the English non-commissioned officer is an excellent man in his way. There is, on duty, less swearing and coarse language in the English army than in any other; on the other hand, punishment is so much the more certain to be applied. The non-commissioned imitates the commissioned officer, and thus [has] adopted manners far superior to those of our German sergeants. Then he does not serve because of the prospect of some pettifogging office in the civil service being held out to him, as is the case with us; he has engaged himself voluntarily for twelve years, and promotion, up to the rank of sergeant-major even, offers him considerable fresh advantages at every step; in every battalion one or two commissions (adjutant and paymaster) are mostly reserved to old non-commissioned officers; and, on active service, every sergeant may attach the golden star to his collar by distinguishing himself before the enemy. The drill-sergeants belonging to this class of men have, indeed, upon the whole, made the volunteers what it was possible to make them in so short a time; they have not only made them steady in company movements, but also licked the officers into shape.

In the meantime, the single companies, at least in the large towns, formed themselves into battalions and received adjutants from the regular troops. Similar to the Austrian, the English subaltern is far less theoretically educated than the North German; but same as the Austrian, if he likes his profession, he knows his duty exceedingly well. Among the adjutants who have passed over from the line to the volunteers, there are men who, as instructors, could not be better; and the results which they obtained in a very short time in their battalions are surprising indeed. Up to the present time, however, only a minority of the volunteers have been formed into permanent battalions, and as a matter of course, these are considerably superior to the mass of companies not so formed.

The volunteers of Lancashire and Cheshire had organized a Review at Newton, half way between Manchester and Liverpool, for the 11th of August, the commanding General of the district, Sir George Wetherall, taking the command. The volunteers who met here were the contingents of the manufacturing districts around Manchester; there were not very many present either from Liverpool or from the neighbouring agricultural districts of Cheshire. To judge from our own German recruiting experience, these corps must have been physically below the average; but it is not to be forgotten that by far the minority of the volunteers belong to the working classes.

The soil of Newton race-course, of itself spongy enough, had been considerably softened by the continuous rains; it was very uneven and very sticky. On one side of it there is a small brook, with here and there some thick gorse on its banks. The ground was just right for a parade of young volunteers; they most of them stood ankle deep in water and mud, and the officers' horses often sank into the clay until above the fetlock-joint.

The 57 corps which had sent in their adhesion were divided into four brigades—the first of four, the remainder of three battalions each; every battalion of eight companies. Lieutenant-Colonels of the line commanded the brigades; officers of volunteers were appointed to the battalions. The first brigade had three battalions deployed, the fourth in column behind the centre. The three remaining brigades stood in second line, nine battalions in continuous columns of companies at quarter distance, right in front.

After saluting the general, a change of front to the left was to be effected, under shelter of the battalion which stood in column behind the first line. To effect this, the two centre companies of the battalion deployed in front of it, wheeled outwards, upon which the column passed through the opening thus formed, and then extended along the water-course—four companies skirmishing, and four forming the supports. The ground and the gorse were both so wet that the men could not be expected to take a correct advantage of the ground; besides, most battalions of volunteers are still occupied with the A B C only of skirmishing and outpost duty, so that it would not be fair to measure them by too high a standard in this respect. In the meantime, the deployed line effected its change of front around its own centre as a pivot; the two centre companies of the middle battalions wheeled a quarter of a circle—the one forwards, the other backwards—after which the remaining companies took up the new alignment. The two battalions on the wings of the first line formed columns at quarter distance, marched into the alignment, and deployed again. It may be imagined what a time was occupied by this complicated and rather clumsy manœuvre. At the same time, the right battalion of the line of columns advanced straight on until halted behind the new right wing of the first line; the remaining battalions faced to the right and followed in double files (fours right), each battalion turning to the front, and following the right battalion as soon as arrived on the spot originally occupied by this right battalion. When the last column had thus arrived upon the new alignment, each column independently wheeled to the left, and thus restored the front of the line of columns.

The third brigade now advanced from the centre of this line of columns; arrived about two hundred paces behind the first or deployed line, the three battalions opened out to deploying distance and deployed in their turn. The chain of skirmishers, in the meantime, having gained considerable ground, both deployed lines advanced a couple of hundred paces, upon which the first line was relieved by the second. This is effected by the first line forming fours right, and the head of each company disengaging and wheeling to the right; files in the second line give way, thus affording room for the first line to pass through; after which, companies form front and wheel into line. This is one of those drill-ground movements which are superfluous wherever they are practicable, and which are not practicable where they would be necessary. After this, the four brigades were drawn together again into a mass of columns, and the troops marched past the general in open column of companies (25 to 35 files front).

We shall not attempt to criticize this system of evolutions which, no doubt, will appear rather old-fashioned to our readers. It is evident that, whatever may be its value in an army of the line with twelve years' service, it is certainly less adapted than any other for volunteers who can afford a few spare hours per week only for their drill. What interests us most on this occasion, is the manner in which these movements were performed by the volunteers; and here we must say that, although there was a slight hitch here and there, upon the whole, these evolutions were gone through steadily and without confusion. The most defective parts were, the wheeling in column and the deployments, which latter were done very slowly; in both evolutions, it was visible that the officers were not sufficiently formed and not yet at home in their duty. But on the other hand, the advance in line, this chief and cardinal movement of British tactics, was good beyond all expectation; the English appear, indeed, to have quite an exceptional talent for this movement, and to learn it uncommonly quick. The marching-past also came off, upon the whole, very well,—and what was most amusing, it came off under a drenching shower of rain. There were a few mistakes against British military etiquette, and besides, by the fault of the officers, distances were very badly kept.

Besides a sham fight organized in London, by some over-sanguine commanders of Volunteers, and gone through rather wildly, this was the first time that a larger body of volunteers performed evolutions, which had something more in view than eventual marching-past. If we consider that the great mass of the troops present at Newton consisted of corps which, counting one, two, or at the outside three companies, are not formed into permanent battalions, have no officers from the regulars, have been drilled by drill-sergeants alone, and have only now and then been brigaded together in a battalion, we shall have to allow that the volunteers have done everything that was possible, and that they are no longer on the same level with our civic guards. As a matter of course, the corps which formed permanent battalions, and are directed by adjutants from the line (for the adjutants, so far, are the virtual commanders of battalions), were also those which went most steadily through their evolutions at the review.

The men upon the whole looked well. There were, indeed, some companies as puny as Frenchmen, but others surpassed in stature the average of the present British line. Mostly, however, they were very unequal in size and breadth of chest. The pallor peculiar to the inhabitants of towns gave to most of them a rather unpleasantly unwarlike look, but eight days' encampment would soon get the better of that. The uniforms, some of them a little over-ornamental, made a very good effect in the mass.

The first year's drill has taught the volunteers so much of the elementary movements, that they may now enter upon skirmishing and rifle practice. They will be far more handy at both these kinds of work than the English line, so that by Summer 1861 they would form a very useful army, if only their officers knew more about their business.

This is the weak point of the whole formation. Officers cannot be manufactured in the same time and with the same means as privates. Up to now it has been proved that the willingness and the zeal of the mass may be relied upon, as far as is required, for making every man a soldier as far as necessary. But this is not sufficient for the officers. As we have seen, even for simple battalion movements, wheeling in column, deployments, keeping distances (so important in the English system of evolutions, where open columns are very often employed), the officers are not by far sufficiently formed. What is to become of them on outpost and skirmishing duty, where judgment of ground is everything, and where so many other difficult matters are to be taken into consideration? How can such men be entrusted with the duty of taking care of the safety of an army on the march? Government has made it binding upon every officer of volunteers to go to Hythe for three weeks, at least. So far, so good; but that will neither teach him to conduct a patrol, nor to command a picket. And yet, the volunteers are chiefly to be used for light-infantry service—for that very kind of duty which requires the cleverest and most reliable of officers.

If the whole movement is to lead to something, this is the point where government will have to step in. All companies which are still existing—singly, or by twos and threes—ought to be compelled to combine together in permanent battalions, [and] to engage adjutants from the regulars. These adjutants should be bound to give to all the officers of their respective battalions a regular course of instruction in elementary tactics, light infantry service in all its branches, and the regulations affecting the internal routine of service in a battalion. The officers should be bound, besides attending Hythe, to do duty, for at least three weeks, with a regiment of the line or militia in some encampment; and, finally, they should, after a certain time, be all made to pass an examination, proving that they have learnt at least the most indispensable part of their business. Such a course of instruction and examination of the officers; further, a medical examination of the men, in order to weed out those who are physically unfit for field-service (and there is not a few); and an annual revision of the company-lists, for the removal of those men who do not attend drill, who only play at soldiers and will not learn their duty—if this was done, the 120,000 men now existing on paper would be considerably reduced, but you would have an army worth three times the one which now counts 120,000 men on paper.

Instead of that, it is reported that the military authorities are busy discussing the important question, whether it would not be desirable to clothe, at the first opportunity, all rifle volunteers in the so very desirable brick colour of the line.

2. Brighton and Wimbledon[1]

The performances of the volunteer forces of London and neighbourhood on Easter Monday appear to have fully born out our anticipations expressed in the article on 'Volunteer Generals'. The attempt of Lord Ranelagh to gather for a day, under his own command, all the volunteers of his district at once created a split among the different corps. An opposition candidate for the commandership-in-chief started up in the person of Lord Bury; to the sham fight at Brighton he opposed a field day at

[[1] *Volunteer Journal*, Vol. II, No. 31, pp. 53–4, April 6, 1861. The Editor of the *Volunteer Journal* stated in a footnote that the article was 'by the author of *Essays addressed to Volunteers*, whose contributions to the *Journal* in future, will be distinguished by the initials placed at the end of the present article'. The initials were 'F.E.']

Wimbledon. Great was the division among the various corps; and the consequence was, that some went to Brighton under Lord Ranelagh, some to Wimbledon under Lord Bury, some to the same place (but independently), some to Richmond, and some to Wanstead. There would be no harm in this dispersion alone. Every corps is quite independent of the other, and has a right to enjoy its holiday after its own fashion. But there must arise, and has arisen, a great deal of harm from the acrimonious debates, the personal bickerings, and animosities which have preceded this split, and which are sure to continue for some time. Commanding officers have taken their post for one side or the other; their men have equally taken part, and not always with their commanders; so that the majority of the London volunteers are broken up into two great parties—the Ranelagh and the Bury factions. At Brighton, a great many men of the corps which had been ordered to Wimbledon appeared without arms, but in uniform to protest against the decision and order of their own immediate superiors; and Lord Ranelagh, enjoying this mark of sympathy amazingly, had them even formed into a provisional battalion, and, with an exquisite military taste not hitherto met with in any army, allowed them to march past with his own men. So, at least, reports the *Daily Telegraph*.

Now, we ask, what right have either Lord Ranelagh or Lord Bury to put themselves forward as candidates for volunteer generals, and thereby to cause dissensions among bodies hitherto acting harmoniously together? Both these officers have served in the regulars; if they had the ambition to become generals, there was for them, as for others, the usual way of aspiring to that position; and, from their social position, they stood a chance ten times better than the great bulk of their other comrades. They knew very well, when entering the volunteers, that the highest active rank compatible with that service is that of lieutenant-colonel; that in case the volunteers were ever called out to act, they would be brigaded together with the line and militia, and placed under the command of brigadiers from the line; that the very nature of the British military organization renders it impossible to appoint general officers from any other branch of the forces than the line. In aspiring to the position of temporary volunteer generals, they aspire to places which neither they nor any other volunteer officers will ever be called on to fill, and which they, from want of experience in the handling of masses of troops, must be incapable of filling. But if, in order to play the general for a day, they disturb the harmony between the various corps of their district, and risk to do the movement serious harm, they deserve even stronger and more unequivocal condemnation.

In all large gatherings of volunteers, hitherto, it has been the usage of offering the commander-in-chief and the appointment of brigadiers and divisionary generals, to the military commander of the district. We have said in our previous article that we fully approve of this proceeding, because it is in accordance with military etiquette and subordination, and because it ensures efficient commanders. Now we see that it does more. Had the command of the Easter performance been entrusted to the proper authorities, there would have been no split, and all this bickering would have been saved. But the London commanders appear to have imbued their men with a highly ludicrous fear of the Horse Guards. 'For God's sake, keep the Horse Guards out!' is their cry. We in the north have not been so particular. We have always been on capital terms with our natural military superiors, and have found the benefit of it; we hope too, that the old system may be continued, and save us from these ridiculous quarrels, now dividing the London force.

How jealous the Londoners were of the Horse Guards is shown by the uproar created by the presence at Brighton of General Scarlett, who was deputed by the Horse Guards to report upon the proceedings. The wise men of the different corps shook their heads in the most serious manner. To send that general here was an attempt on the part of the Horse Guards to put in the thin end of the wedge. The most fearful consequences were predicted if this were allowed to pass as a matter of course. The volunteers ought to protest; and, indeed, it was proposed that General Scarlett was not entitled to the salute which was due to the lord-lieutenant of the county only. The matter was finally settled by both coming up and receiving the salute in common. But that such questions could be discussed, shows how much some volunteers do mistake their position.

Thus we see that neither as regards discipline within the corps, nor subordination or even deference to superior officers, has this Easter affair been of any benefit to the London volunteers.

In turning to the various field days, we must promise that we can only go by the reports of the London press, which are exceedingly incomplete and obscure as to military features; and if we should make mistakes in facts, it cannot, therefore, be laid to our charge.

Lord Ranelagh's five brigades took up a position east of Brighton, facing the town, after having marched past. They were very small, each numbering three battalions of 400 men on an average. With this force a ridge of hills was to be occupied, which was far too extensive for such a small number. Now, in this case, if 7,000 men accept a combat, the supposition is, that the enemy is not of a very great superiority in numbers, as otherwise they would retire on their reserves. Consequently, the commander would form his troops in a first and second line, and a reserve, as usual; supporting his flank as best he could, and trusting to his reserves and to the main body (supposed to be in his rear) for the repulse of any outflanking movements on the part of the enemy. But as it would appear by almost all reports, Lord Ranelagh extended the whole of his 7,000 men in one single line! He had a programme made out for three times that number, and as only 7,000 had come instead of 20,000, he made the small number occupy the whole extent of ground marked out for the expected larger number. If this has been actually done, it would settle at once and for ever Lord Ranelagh's claims to generalship, volunteer or other. We are most unwilling to believe that he should have committed such an absurdity, but we have never seen the almost unanimous statement of the press contradicted, and, therefore, must believe it to be the case. We are even told that there *was* a small reserve of a few companies, but that two-thirds of it were at once called into the first line, so that scarcely the ghost of a second line, or reserve, was on the field.

This first line, with its *supposed* second line and *supposed* reserve, was attacked by a supposed enemy who was received by skirmishers, and after these had been thrown back, by file-firing from the right of companies. Why the volunteers are taught file-firing in sham-fights is more than we can tell. We believe that all soldiers who have seen service will agree with us that file-firing, of some use at the time when the lines advanced at the goose-step, is now completely antiquated, that it never can be of any good in front of the enemy, and that there is no useful intermediate link between the fire of skirmishers and the volley.

The imaginary enemy repulsed the defensive line. How the action of second line and reserves (which must, after all, have been *supposed* to support the first line) was represented, we are at a loss to understand. The battalions had to suppose, not only that they were repulsed, but also that they were relieved. A second line of hills to the rear was then occupied and lost, but at a third accident of ground matters took a turn, and imaginary reinforcements coming up, the enemy was beaten back but not seriously pursued.

We are told by *The Times* that the movements gone through were of the simplest nature. The following is a summary which the correspondent of the *Telegraph* got from an officer, as a report of the movements of his battalion: 'Having arrived in fours, the ranks formed a quarter-distance column in front of No. 1; column wheeled to the left and deployed again on No. 1, advanced in line, covered by No. 1, halted, the assembly was sounded and the skirmishers came in; firing from the right of companies, line retired, and from the proper right of companies passed by fours to the rear, front turn into column, formed quarter-distance column in rear of No. 1, marched by sub-divisions round the centre, opened out to wheeling distance from the rear; left wheeled into line and fired a volley, moved in column of companies from the right along the rear, lined to the front on No. 1, formed quarter-distance column in front of No. 1; deployed on No. 2; then No. 1 advanced to the front, and the remainder right wheeled; formed quarter-distance column in rear of No. 1; fours left, and so left the hill.' Of the way in which these movements were carried out, we only know that, as usual with volunteers, distances very often were lost, and companies got asunder when forming line.

At Wimbledon, Lord Grosvenor manœuvred his battalion early in the morning, and marched off when Lord Bury's two brigades (under 4,000) arrived. These went through a very simple performance, but very well adapted to give the men an idea of events and evolutions such as will occur in actual war. The whole is so well described in Colonel McMurdo's address that we have merely to add that here, too, we find file-firing used to fill up the interval between the retreat of the skirmishers and the opening of volley-firing—a thing we most decidedly consider faulty in every respect. The Duke of Wellington rather let his men lie flat down in such moments than stand up to be shot at by artillery and return a weak, ineffectual, and, to themselves, demoralizing file-fire.

For the remainder, we concur entirely with Colonel McMurdo's admirable address, with which we conclude these remarks. We hope all volunteers will note and bear in mind what he says on company drill. The elementary instruction of volunteers must necessarily be less perfect than that of the regular soldier, but it is nevertheless of the highest importance in giving solidity to battalions. The greatest attention to company drill alone can make up in some degree for this unavoidable defect.

Colonel McMurdo says:

'Volunteers, to men of understanding it is not necessary that the movements which you have gone through today should be fully explained, but I think it necessary to call your attention to the nature of the two positions which you have taken up in the course of the field movements through which you have gone. The first position which you took up was naturally one of very great strength—so great that two-thirds of the enemy would have been non-effective. His cavalry could not have acted with effect, nor could his artillery have injured you except by a vertical fire. It was supposed that the enemy, finding that too strong a position, endeavoured to reach the plateau on which we now stand by turning our flank up one of those long valleys in the direction of Wimbledon. It was necessary, therefore, that you should quit the strong position which you formerly held by changing your front to the left. The enemy had a double object in view. He desired to come upon plain level ground, by which means he could bring both his artillery and his cavalry to bear in the action, as well as his infantry; he also desired by turning your left flank, to reach the Wimbledon Road, by which he could march through you on London. It is my desire to point out to you the difference in the two positions which you held. It was a very different thing when you were along that rugged crest of a hill, where neither cavalry nor artillery could reach you. You checked the enemy there, and any number of bold men could have checked an enemy there; but here you are brought, as it were, on a sort of billiard table, where you might be exposed to the operations of, perhaps, the best troops in Europe. I observed, in forming the line here, that some battalions were a little unsteady. I do not blame them for it, because they have as yet had very little practice. Still they were unsteady; and if they were unsteady in coming into line to-day, what would be the case if this plain were swept by the artillery of an enemy, if you were choking with thirst, many of your comrades falling around you, and suddenly, through the dust and smoke, you felt the very ground shake under you by a heavy charge of the enemy's cavalry. Consider how liable young troops would be to be unsteady under such circumstances. What is it that overcomes all this? It is discipline, and discipline alone. By the term discipline I do not mean the correction of bad conduct—I mean that habitual union, that combination of mind and body brought to bear upon a certain object; that combination of mind and body that sets the whole in action, and makes a company, a battalion, or a brigade, act like a machine. Now this can only be acquired by company drill; it can only be acquired by paying great attention to individual drill, because I consider a company to be the unit of an army, and when individuals are well drilled and steady the company is steady, and the whole army will be steady. All that you have learned in the way of shooting, all your zeal, all your patriotism, will be of no avail in the day of battle without a thorough knowledge of your company drill. Company drill and nothing but company drill will do, and therefore I beg you to consider that excellence in shooting is not everything, because nothing will do unless you have perfect steadiness of formation under fire. Gentlemen, you have had a hard day's work on the wet ground, and therefore I will no longer detain you, but leave you to return to those homes which you are so well able to protect.'

3. A Military Criticism of the Newton Review[1]

Last year's Newton review was a great success, the greater because beset by difficulties of all kinds. It was the first attempt to bring together the volunteers of Lancashire in a body; the

[[1] *Volunteer Journal*, Vol. II, No. 49, p. 199, August 10, 1861. Article by the author of 'A German Account of the Newton Review, 1860.']

railway arrangements were anything but what they should have been; the ground was in an execrable state; the weather was very bad. In spite of all this, the thing went off uncommonly well, and our volunteers went home, wet, hungry, thirsty, but with the proud consciousness that they had surprised everybody by the cool, steady and soldierlike manner in which they had gone through their work.

Can as much be said of this year's review? We are afraid it cannot. The railway arrangements were excellent; the ground was in capital order; the weather was fine; the volunteers had gone through another year's drill; and yet, we are sure, most of them went home less satisfied with their day's work and their day's success than last year. Whose fault was that?

When the troops arrived on the ground, the flags marking out the positions of the various brigades were in their places, and generally the battalion aids were at once placed. But a good many of the battalions, especially those which arrived first, were moved about, halted, again moved, and again halted for a long time before they were brought to their proper places. The consequence was that corps which were from half an hour to an hour on the ground before the review commenced, could not find time to pile arms and dismiss their men for even a few minutes to get refreshment. This was certainly not the fault of the commanders of battalions.

After the general salute, the evolutions commenced. But there were scarcely any evolutions. The first brigade deployed, and went through a series of firing one round by companies from centre to flank, one volley by battalions, three rounds file-firing. In the meantime the second brigade deployed, and after the firing was over, relieved the first line. This was done by both lines forming fours deep, and the fours of the second line passing through the spaces of the first. The regulations themselves characterize this movement as one adapted for *parade purposes* only, and never to be applied on service (p. 113). Then the second brigade went through the same course of firing, while the third brigade deployed to form a second line, and the first brigade fell back to the rear in column. We noticed that the first brigade was a very long while over this, and only got out of the way when the firing of the second brigade was nearly over. Then the third, and afterwards the fourth, brigade advanced and took their turn of firing, after which the whole body formed in mass of columns and marched past.

Thus, it is evident, instead of evolutions, there were but two points in which the volunteers present could show their proficiency—the firing and the marching past. Now, we protest against blank cartridge firing being made a test by which to judge such a body as the volunteers assembled at Newton. There were regiments which have fired immense numbers of blank cartridges, and which, consequently, long ago obtained considerable success in sharp, round volleys. There were others which are quite as well, perhaps better, up in their company and battalion drill and in target practice, but which scarcely ever fired blank cartridges before. And there were a great number of the small country corps, formed into battalions for the occasion, which never had a chance of firing a battalion volley, for the very simple reason that, so far, they have not been able to go even through battalion drill. Volley-firing, as far as it is to be judged by the sound only, and not by the effect, is of all the duties of a soldier by far the easiest; an otherwise steady battalion will learn it in a very short time, and if the great majority of the battalions present gave very bad volleys, indeed, we must say we are more pleased with it than otherwise, inasmuch as it shows that they have not wasted their time with practising an art which they can learn in a week at any time, and which is very apt to be indulged in as a plaything and an advertisement.

The only good point in the programme was that it gave the whole of the infantry present something to do. Otherwise it was very poor indeed. There was no skirmishing, scarcely any evolutions, and there was a test of efficiency set up which was not only delusive, but positively unfair to the mass of the corps present. As to the gallant cavalry charge which concluded the manœuvres, we [had] better say nothing of it. The public took it for a capital joke.

In the marching past we noticed again the everlasting weakness of volunteers—the utter disregard to distances. Only one regiment came past with anything like the proper distances, and it was *not* the one which had distinguished itself so much by its volleys. Now, we think that proper distance-keeping is both more difficult and more important, in the present style of volunteer drill, than sharp volleys. Upon the whole, the marching past showed less improvement upon last year than one had a right to expect; but we are bound to say that in this respect the smaller corps from the country had made the greatest progress. This deserves so much the more a public acknowledgment as these small corps have to struggle against the greatest difficulties, are mostly deprived of the assistance of adjutants, and have no higher military authority to go by than their drill-sergeant.

We noticed with regret the progress of the scarlet coat, and even the bear-skin cap, among the Lancashire volunteers; it seems to indicate a hankering after show, which cannot do the movement any good. This is a subject, however, which would bear us too far away from Newton, and we shall, therefore, return to it on some other occasion.

4. Volunteer Artillery[1]

We give, in another column of this week's journal, some remarks from the London correspondent of the *Manchester Weekly Express*, on volunteer artillery. The subject is one of great importance, and ought to be widely discussed; the more so, as the part which the volunteer artillery is to take in the defence of the country does not appear to have been, as yet, very clearly defined. The very article to which we refer, while it wishes to see the formation of artillery corps confined to the sea-board, still expects volunteer gunners to act as a kind of field-artillery, not confining themselves to the attendance upon heavy guns in fortified places, but also galloping about with 'light six-pounders or Whitworth's twelves'.

Now, it is evident that the first question to be settled is the proper sphere of action of the volunteer artillery. Unless this be done, there will never be any uniform system of training in the different corps; and as the science of artillery comprises the most multifarious subjects, the whole of which it would be difficult indeed, theoretically and practically to teach to all the volunteer officers and privates, the different corps, when wanted for action, would arrive with very different qualifications for the duties to be performed by them; and many a company, on being put to a particular task, would be found to be very little qualified to carry it out.

[[1] *Volunteer Journal*, Vol. I, No. 6, pp. 87–8, October 13, 1860.]

In the following observations, we do not by any means profess to say what volunteer artillery ought, or ought not, to be; we merely wish to point out some of the conditions under which volunteer, as well as any other artillery, has to be formed, to open the field for that discussion which we invite, and from which, ultimately, an understanding must arise, as to the proper sphere of action of volunteer artillery corps.

All artillery is divided into field artillery which has to operate with the infantry and cavalry in the field, and is provided with horsed guns; and in siege or fortress artillery which works heavy guns in stationary or protected batteries, for the attack or defence of fortified places. If in a regular army, the length of service of the men, and the special scientific education of the officers, renders it possible to train the whole body to both branches of the service, so far, at least, that on an emergency every company can be put to any duty; this is not the case with volunteers who, officers as well as men, can devote but a portion of their time to their military duties. In France, in Austria, in Prussia, field-artillery is kept quite distinct from garrison or siege artillery. If this is the case in regular standing armies, surely there must be some reason for it which will operate far stronger in an army of volunteers.

The fact is this: the mere handling of a field-gun is not so different from that of a heavy gun in battery that the privates of a volunteer company could not easily learn both. But the nature of the duties of the officers in either case are so very different, that nothing less than a professional education and long practice could qualify a man to do both equally well. In an officer of field-artillery, a rapid military glance, a thorough judgment of ground and of distances, a perfect knowledge of the effect of his guns, enabling him to hold out against an attack to the last moment without losing any guns, a long experience of what horses can do, and of the way to treat them in a campaign, and finally, a good deal of dash combined with prudence, are the chief qualities. In an officer of garrison or siege artillery, scientific acquirements, theoretical knowledge of artillery in all its branches, of fortification, mathematics, and mechanics, an ability of turning everything into use, a patient and strict attention to the erection and repair of earthworks, and to the effects of a concentrated fire, and a courage more tenacious than dashing will be required.

Give the command of a bastion to a captain of a 9-pounder battery, and it will take the best man a deal of training before he is up to the work; put an officer who has attended for a couple of years to nothing but siege guns, at head of a battery of horse artillery, and it will take a long while before he has worn off his methodical slowness and recovered the dash required for his new arm. With non-commissioned officers lacking the scientific education of their superiors, the difficulty will be still greater.

Of the two, the garrison artillerist seems to be the easiest formed. Civil engineers possess all the preliminary scientific knowledge required for the business, and will very soon learn the application to artillery of the scientific principles with which they are conversant. They will easily learn the handling of the different machines used in moving heavy ordnance, the construction of batteries, and the rules of fortification. They will, therefore, form the class from which volunteer artillery officers should be chiefly selected, and will be especially adapted for garrison artillery. It will be the same with the non-commissioned officers and gunners. All men who have had much to do with machinery, such as engineers, mechanics, blacksmiths, will form the best material, and on this ground, the great manufacturing centres ought to form the best corps. Practice with heavy guns may be an impossibility in the interior of the country, but the sea is not so very far from our Lancashire and Yorkshire inland towns that occasional trips to the sea-side might not be organized for the purpose; besides, with heavy guns in battery, where the first graze of every shot can be seen, and the men can correct themselves, actual target practice is not of such paramount importance.

There is another thing against the attempt at getting up volunteer field-artillery—the expense of the guns and the horsing of them. A few companies combining amongst themselves may, indeed, be able to raise the expense of horsing a couple of guns for the summer months, and drill with them in turns, but neither men nor officers will thereby be formed into efficient field-artillerists. The expense of equipping a field-battery of six guns is generally reckoned about equal to that of getting up a whole battalion of infantry; no company of volunteer artillery could afford such an outlay; and considering the disgrace attached to the loss of a gun on the battlefield, it may well be doubted whether any government would ever be inclined, in case of invasion, to entrust volunteer artillery with field-guns, horses, and drivers, on the terms on which rifle volunteers are supplied with small-arms.

On these and other grounds, we cannot but come to the conclusion that the proper sphere for the volunteer artillery is the manning of heavy guns in stationary batteries on the coast. An attempt at field-artillery may be inevitable in inland towns, to keep up the interest in the movement, and it will certainly do no harm to either officers or men to be made acquainted, as far as possible, with the handling of horsed light guns; but we confess we have, from our own personal experience in the arm, our great doubts as to their eventual proficiency in field-service. Still, they will have learned a great many things which will be quite as useful to them in the use of heavy guns, and they will soon be up to the mark when placed in charge of them.

There is another point we wish to allude to. Artillery, far more than infantry and cavalry, is an essentially scientific arm, and as such, its efficiency will chiefly depend upon the theoretical and practical knowledge of the officers. We have no doubt that by this time Major Griffiths's *Artillerists' Manual*[1] will be in the hands of every officer of volunteer artillery. The contents of that book show with what a variety of subjects an artillery officer, and even a non-commissioned officer has to make himself familiar before he can lay claim to any proficiency in his arm; yet, that book is merely a short abstract of what an efficient artillerist ought to know. Besides the regular company and battalion drill common to infantry and artillery, there is the knowledge of the many different calibres of ordnance, their carriages and platforms, charges, ranges, and various projectiles; there is the construction of batteries, and the science of sieges; permanent and field fortification; the manufacture of ammunition and fireworks; and, finally, that science of gunnery which, at the present moment, is receiving such wonderful and new additions by the introduction of rifled guns. All these things have to be learnt both theoretically and practically, and they are all of equal importance; for whenever the volunteer artillery are embodied for active service, they will come to a deadlock unless all these

[[1] F. A. Griffiths, *The Artillerist's Manual* . . . (Woolwich, 1st ed., 1839, 10th ed., 1868.]

branches have been attended to. Of all volunteer corps, therefore, the artillery is the one in which the efficiency of the officers is of the greatest importance; and we do hope and trust that they will exert themselves to the utmost to attain that practical experience and theoretical knowledge without which they must be found wanting on the day of trial.

5. Volunteer Engineers[1]

The volunteer army has had, for some time, its infantry and artillery in considerable numbers; it has had its small complement of cavalry too; and now, the last branch of military service, the engineering branch, is gradually being taken up. The subject of volunteer engineers is at present very widely discussed, and it deserves the attention it enjoys. The corps of Royal Engineers is too weak already for the numerous duties it has to perform at home and in the colonies. What will it be in case of a war, and anticipated invasion? Then the numerous fortifications which now are in course of erection, and by means of which the dockyards are being surrounded by vast entrenched camps, will require a considerable number of engineer officers and men for their garrison; and the army in the field, swelled to twice or three times its present number by the addition of the volunteers, will also be in want of a certain complement of engineers, to give it its full liberty of action before the enemy. Unless the corps of Royal Engineers is considerably increased, the duties of this branch of the service must either be imperfectly performed, or they must be performed by volunteers trained for them beforehand.

The number of engineers to be attached to an army in the field is, after all, not very numerous; three or four companies to an army corps of two divisions (16 to 24 battalions of infantry, with a due proportion of cavalry and artillery) would be quite sufficient. Supposing a field-army of 40,000 of the line, 20,000 militia, and 100,000 volunteers, in all 160,000, or 200 battalions, this would give from eight to ten corps, and require about thirty companies of engineers. We will suppose ten companies to be furnished by the Royal Engineers; this would leave twenty companies to be supplied by the volunteer movement. About the same number more volunteer engineers would be sufficient to assist the Royals

[[1] *Volunteer Journal*, Vol. I, No. 12, p. 163, November 24, 1860.]

in the defence of the fortified dockyards; so that something like forty companies of volunteer engineers would appear an ample complement for the present strength of the volunteer infantry and artillery. If the number of volunteers should so far increase as to enable them to appear in the field, after deducting garrisons, with more than 100,000 men, one additional engineer for every hundred additional riflemen would be enough; giving 200 engineers (or three companies) for every army corps of 20,000 men.

For the present, then, forty companies, or about 3,000 effectives, would be the maximum engineer force which it might be advisable to create. And it will require a great deal of energy to make them engineers not only in name, but also in reality. We find already now that among artillery volunteers a great deal of time is devoted to company and battalion drill, carbine in hand, although all this work serves for parade purposes only, and will never avail them one jot on active service, be it with field-guns, or be it in fortifications. And we are afraid it will be the same with the engineers. They should, above all things, bear in mind that every hour spent on company drill, beyond what is required to give them a military bearing, a ready and instantaneous obedience to orders, and the capability of moving in good order on a march, is an hour lost to them; that they have quite different things to learn, and that on these, and not on steady marching past, depends their efficiency. They will have to acquaint themselves—men as well as officers—with the elements of field and permanent fortification; they will have to practise the construction of trenches and batteries, and the making and repairing of roads. If means can be found, they will have to construct military bridges, and even to dig mines. Some of these branches, it is to be feared, can only be taught theoretically, as fortresses in England are scarce, and pontoons also; and not every volunteer can be expected to go to Portsmouth or Chatham to study fortification or assist at the laying down of a pontoon bridge. But there are others which it is in the power of every company to practise. If there was a company of engineers formed here in Manchester, we could show them plenty of lanes in as bad a state as any to be passed by a column in war, and where those whom it concerns would very likely be only too glad to allow them to practise road-making to their heart's content. It would not be very difficult for them to find a plot of land on which they could construct a few field-works, dig trenches, and erect batteries; especially as such a plot of land would offer both the artillery and rifle volunteers an opportunity of practising such parts of their service as they could otherwise not be made to go through. They might even find spots where they would be allowed occasionally to throw a small bridge of *chevalets* over one of those high-banked rivers of our neighbourhood which offer such capital facilities for this kind of bridge wherever their bottom is firm. Such things, and many others of the same kind, should constitute their chief practice; company drill should be gone through rapidly at first, and only taken up again when the corps have got on fairly with their real engineering business; then, in the second winter, the nights may be used for drill with advantage. But if the engineers make it a point, from the beginning, to compete with the rifles in the style of marching past, and in battalion evolutions, to the detriment of their specific education; if the attention of the officers is directed more towards the duties of an infantry officer than to professional education—then the volunteer engineers may depend upon it that in a campaign they will far oftener be used as infantry than as engineers.

There will be little difficulty in finding very efficient officers, if they are selected from the only class fit for the post—the civil engineers. A few months' theoretical study, and an occasional journey to Chatham, Portsmouth, or Aldershot, will soon make them conversant with most branches of military engineering, and the military education of their companies will help them on. They will learn by teaching. Their own profession compels them to know all the principles of military engineering, and as they *must* be very intelligent and well-informed men, the application of these principles to military subjects will give them but little difficulty.

We have read a statement in the *Army and Navy Gazette* respecting some immense military engineering organization, which is to compromise all the lines of railway in the country, and to promise vast results in case of an invasion. The principal features of the plan are reproduced in last week's *Volunteer Journal.* The shape in which this plan is presented before the public is excessively vague; so far, we do not see the immense advantages that are ascribed to it, and rather think that two different things have been mixed up together. No doubt it is of the highest importance to study the strategical bearings of every single line of railway in the kingdom, as well as of the whole network of railways combined. This is so important that we should consider it a grave delinquency if it had not been done long ago, and if there were not now lying in the archives of the Horse Guards, as well as of the various district commanders, very extensive papers embodying the results of these studies. But this is the duty of the staff, and not of the engineers. As to forming the engineers, firemen, platelayers, and navvies of every railway line into a corps of military engineers, we do not see the great advantage of this. These men have already, so to say, a military organization, and are under stricter discipline than any volunteer corps in the country. What they are expected to do in their quality as volunteer engineers, they are quite as capable of doing in their present capacity. And as in time of war their presence at their present posts would be far more indispensable than now, there can be no earthly use in training them to special branches of military engineering.

These remarks apply to the plan only as far as it has been made public. If it should turn out, hereafter, that it contains other features, we must, of course, reserve our opinion.

6. Volunteer Generals[1]

There has been one thing wanting to the volunteer movement, and that is a fair and intelligent, but plain and outspoken criticism by competent outsiders. The volunteers have been to such a degree the pets of the public and the press, that such a criticism became an absolute impossibility. Nobody would have listened to it; everybody would have declared it unfair, ungenerous, untimely. The shortcomings of volunteer performances were almost invariably passed over in silence, while every corps was extolled to the skies for whatever it did go through tolerably well. The politeness of people, with any regard for impartiality, was most fearfully taxed; everywhere they had to give their opinion upon some volunteer affair or other, and unless they were prepared to utter the most fulsome and unqualified praise, they were lucky if they escaped being thought conceited snobs. How often have the volunteers been insulted by the stupid piece of flattery that they were fit to fight any troops in the world? How often have they been told that no division of the line could have done

[[1] *Volunteer Journal*, Vol. II, No. 28, p. 29. March 16, 1861. Article by the author of *Essays addressed to Volunteers*.]

better what they did at Hyde Park, Edinburgh, Newton, or Knowsley?

Now, setting aside such absurd flattery, which at all times would have been ridiculous, we are quite prepared to admit that a fair trial had to be given to the volunteers before a fair opinion could be passed on their proficiency. But that time has passed long ago. If the volunteer movement, after nearly two years' existence, cannot yet bear criticism, it will never be able to bear it. The great reviews of last summer, in our opinion, mark the period at which the movement passed from infancy into adolescence; by these reviews the volunteers themselves actually provoked criticism; and yet that criticism, with one or two exceptions, was not publicly exercised by those who ought to have done so.

The effects, as well of this absence of plain and outspoken criticism as of this unmitigated adulation, are now visible enough. There will be scarcely a single volunteer corps of eighteen months' standing which does not consider itself, in the silence of its own conviction, quite as good as it has any business to be. The men, after having gone through the simplest battalion movements, through the routine work of skirmishing on a level piece of ground, and through a little rifle shooting, will be but too apt to say that they can do all these things as well as the line; and what the officers think of themselves has been shown by the race for promotion to captaincies, majorities, and lieutenant-colonelcies, which has been going on in almost every corps. Everybody considered himself perfectly fit for any commission he might be able to procure; and, as in the majority of cases, it was certainly not merit which made the man, we need not wonder that, in a good many instances, we have anything but the right man in the right place. Officers and men so firmly believed in what a benevolent press and public chose to call the perfection of their performances, that they began to think soldiering an uncommonly easy thing; and it is a wonder their own mushroom-perfection did not make them consider a standing army, composed of long-trained officers and soldiers, quite unnecessary in a country where perfect soldiers could be manufactured far easier on the volunteer plan.

The first distinct proof of the damage done to the movement by its friends in the press, was the sham fight last summer in London. Some enterprising colonels of volunteers thought the time had come to give their men a foretaste of what fighting looked like.

Of course, the wiseacres among the regulars shook their heads, but that did not signify. These regulars bore an ill-will to the volunteer movement; they were envious of them; the success of the Hyde Park review almost made them go mad; they feared the sham fight would come off better than anything the line had ever done in that branch, etc. Had not the men gone through the manual and platoon, battalion drill, and skirmishing? And the officers, though mere civilians a short time ago, were they not now efficient captains, majors, and colonels? Why should they not lead a brigade or a division, as well as a battalion? Why should they not play a little at generals, having so well succeeded in the lower grades?

Thus did the sham fight come off, and a regular sham it was, according to all accounts. The thing was gone through with a supreme contempt for all accidents of ground, with a splendid disregard for the effects of fire, and with a perfectly ludicrous exaggeration of all the impossibilities which are inherent in every sham fight. The men learned nothing by it; they took home with them an idea of fighting totally the reverse of reality, an empty stomach, and tired legs: the latter two, perhaps, the only things which might be considered in any way useful to incipient warriors.

Such childishness was pardonable in the boyhood of the movement. But what shall we say to the return of similar attempts at this present time? Our indefatigable London self-made volunteer generals are at work again. Their own laurels of last summer do not let them rest. A mere sham fight on an ordinary scale no longer satisfies their ambition. This time a great decisive action is to be fought. An army of 20,000 volunteers will be thrown from London upon the south coast, will repel an invasion, and return to London the same evening, so as to be able to attend to business next morning. All this, as *The Times* very properly observes, without any organization, without staff, commissariat, land transport, regimental train—nay, without knapsacks, and without all those necessaries for campaigning which a line soldier carries in that receptacle! However, this is but one side of the question; it shows only one striking feature of the incredible self-confidence which our volunteer generals have the satisfaction of possessing. How the mere tactical knowledge, the art of handling the troops, is to be procured, *The Times* does not inquire. Yet this is quite as important a point. The drill of volunteers, so far, has been gone through on level ground only; but battle-fields generally are anything but level and unbroken, and it is just the taking advantage of this broken and undulating ground which forms the basis of all practical tactics of the whole art of disposing troops in action. Now, this art, which has to be learned theoretically and practically, how are the volunteer generals, colonels, and captains to know it? Where have they been taught it? So little has this groundwork of practical tactics been attended to, that we do not know of a single corps which has been instructed, practically, in skirmishing in broken ground. What, then, can become of all such attempts at sham fights but a performance, which, satisfactory, perhaps, to ignorant spectators, will be most certainly useless to the men, made to go through it, and which cannot but tend to make the volunteer movement look ridiculous in the eyes of military men assisting at such a spectacle.

To our astonishment we find that even in practical Manchester an attempt is made to manufacture volunteer generals. No doubt we are not quite so advanced as our friends the Cockneys; we are not to have a sham fight, but a mere field-day of all the Manchester volunteers—something, it appears, in the style of the Newton review; and the affair is to come off on some comparative level piece of ground. Now we wish it to be understood that, so far from disapproving this, we think, on the contrary, that half-a-dozen such field-days every year would do the Manchester volunteers a deal of good. We would add, that we should even consider it desirable that these field-days should come off in ground a little more broken, so as to allow the manœuvres (against a supposed enemy) to come off with more variation, and to gradually give officers and men the habit of manœuvring in broken ground. Such manœuvres would give the adjutants excellent opportunities for afterwards connecting with them, at officers' drill, a few practical lectures on the mode of taking advantage of ground in fighting. So far, then, we not only approve of the plan, but should even wish to see it extended and regularized. But, then, we are informed by a paragraph,which appeared last Saturday in a local paper, that on this occasion the volunteers will do everything for themselves. That is to say, they are going to have a volunteer commander-in-chief, volunteer generals of brigade, and a volunteer staff. Here, then, we have the attempt to import into Manchester, the London system of manufacturing volunteer generals, and to that we decidedly object. With all due respect to the commanding officers of regiments in Manchester, we saw they have yet a great deal to learn before they become—and we make here no exception—fully efficient commanders of battalions; and if, before they have made themselves fully equal to the responsibility already undertaken by them, they aspire to act for a day in higher commands, we say that they do that which would be the greatest curse to the volunteer movement, namely, playing at soldiers, and that they degrade the movement. At the head of their battalions they would be in their places, they would be able to look after their men, and they would learn something themselves. As Brummagem generals, they would be of no real use, neither to their men nor to themselves. All honour to the adjutants of our Manchester regiments, who deserve the greater part of the credit of having made their regiments what they are; but their place is with their respective regiments, where, as yet, they cannot be spared, while they would be of no real use to those regiments if they played, for a day, at adjutant, general, and brigade-major—a thing which surely would not give them, personally, any particular satisfaction.

When we have in Manchester the headquarters of the northern division of the army, with a numerous and efficient staff—when we have an infantry and a cavalry regiment garrisoned here—surely there is no necessity of recurring to such extraordinary pranks. We think it would be both more conformable to military subordination, and also more in the interest of the volunteers themselves, not to collect in such numbers, under arms, without offering the command to the general of the district, and leaving to him the choice of appointing staff and line officers to the division and brigades. No doubt the volunteers would be met in the same friendly spirit as they have been on former occasions. They would then have men at the head of the division and brigades who understand their business, and can point out mistakes when they occur; and they would also preserve their own organization unbroken. No doubt this would preclude colonels from acting as generals, majors as colonels, and captains as majors; but it would have the great advantage of keeping out of Manchester that manufacture of Brummagem generals for which London is now getting an unenviable notoriety.

7. Company Drill[1]

In our last number but one we called the especial attention of volunteers to the remarks of Colonel McMurdo on company drill. We now recur to the subject as we think it is high time that its importance should be fully appreciated by every rifleman in the country.

The other day we took occasion to witness the battalion drill of a volunteer corps, which on the whole, stands decidedly above the average of the force of this district in proportionate number of effectives, good attendance at drill, attention to duty on the part of officers, and, consequently, in general efficiency. To our great surprise, we found that there was very little progress beyond what we had seen this same corps perform some six months ago. The battalion movements came off slightly better than at the close of last season, but the manual and platoon were gone through in a rather slovenly manner. Even in shouldering arms, every man looked as if acting without any consciousness that he was to act in concert with some 400 men right, left, and in rear of him. In making ready and presenting, every rifle seemed to take a pride in coming to the proper position independently of its neighbours; and, altogether, a quiet disregard of the one—two, or one—two—three, by which the execution of each word of command is to be characterized, appeared the general order of the day.

In one corner of the barrack-yard in which this took place, we happened to see a squad of a line regiment fall in for drill under a sergeant. They were, we suppose, the awkward squad of the battalion, ordered for extra drill. What a difference! The men stood like statues; not a limb moved till the word was given, and then those limbs only moved which had to execute the command—the remainder of the body remained perfectly still. When the command struck their ears, every arm moved simultaneously, every motion into which the execution of the command was divided was perfectly distinct, and was gone through at the same moment by every man. The whole squad, in fact, moved like one man. Those gentlemen who are so fond of boasting that the volunteers can do all their work quite as well as the line, would do well to go and study the line a little; they would then soon find

[[1] *Volunteer Journal*, Vol. II., No. 33, pp. 69–70, April 20, 1861.]

out that between the best volunteers and the worst drilled line regiment there is still an enormous difference.

But what, it will be said, is the use of such perfection of drill to the volunteers? They are not intended to have it, they cannot be expected to have it, and they will not require it. No doubt this is quite correct. The very attempt to make volunteers emulate the line in perfection of drill would be the ruin of the movement. But drilled the volunteers must be, and so far drilled that common simultaneous action shall become quite mechanical, quite a matter of course with them; so far, that all their movements and motions can be gone through steadily, simultaneously, by all, and with a certain degree of military bearing. In all these points the line will remain the model which they will have to look up to, and company drill will have to be the means by which the required efficiency can alone be obtained.

Take the manual and platoon. That on any given word of command, the whole of the rifles in the battalion should be moved simultaneously, and in the manner prescribed, is not a mere matter of appearance. We must suppose that all volunteer corps are now so far advanced that the men can go through this exercise without positively hurting each other, or knocking their rifles together. But even beyond this, a mere slovenly way of going through the different motions has, undoubtedly, a great moral effect upon the battalion under drill. Why should any one man be particularly attentive to the command, if he has blunders committed right and left, and rifles coming up or down in a straggling way long after he has performed the command? What confidence, before the enemy, can a man on the left wing have in his comrades on the right wing, unless he knows they will load, make ready, and present together with him on the command being given, and will be ready again, as soon as he himself shall be, either to fire again or to charge? Moreover, every experienced soldier will tell you that the habit of such simultaneous action—the certainty of the officer's command being responded to by those two or three round distinct sounds, denoting that every man acts at the same time as his comrades—has a very great moral influence on the battalion. It brings home to the senses of the men the fact that they really are like one body; that they are perfectly in the hand of the commander, and that he can employ their strength at the shortest notice and with the greatest effect.

Again, take the movements of large or small bodies of troops. Unless every man is so far confirmed in his drill that every movement he may be required to go through is done almost mechanically on the word being given, a battalion will never move steadily. A soldier who has still to ransack his memory or his intellect to make out what kind of thing the command given asks him to do, will do more harm than good in a battalion. So will a man who, either from habit or some other cause, is apt to think that certain movements will necessarily be followed by others; he will often receive a command quite different to what he expected, and then he will very probably blunder. Now, these defects can only be overcome by constant company drill. There the officer in command can put the small body under his orders, in a quarter of an hour's time, through so many different movements and formations, and can vary the order of passing from one to the other to such an extent, that the men, never knowing what is to come, will soon learn to be attentive and to respond quite mechanically to the word of command. In a battalion, all movements are necessarily much slower, and therefore on the whole less instructive to the men, though more so to the officers; but it is an acknowledged fact that men, perfect in their company drill, will, under good officers, learn their battalion movements perfectly in a very short time. The more the men are tossed about in company movements by a competent quick-eyed instructor, the steadier they will afterwards be in the battalion. And it requires no pointing out how important perfect steadiness in a battalion is: a volley may be given rather irregularly, and still take effect; but a battalion thrown into disorder in forming square, deploying, wheeling in column, etc., may at any time be hopelessly lost if in front of an active and intelligent enemy.

Then there is the important point of distances. It is an indispensable fact that no volunteer officer or soldier has an eye for distances. In marching in open or quarter distance column, in deploying, every battalion drill shows how difficult to the officers it is to keep the correct distance. In re-forming column from square, the men of the centre sections almost always lose their distance; they step back too far or too little, and the wheel backwards is consequently done in a very irregular way. The officers can learn to keep distance in the battalion only, though company movements in sub-divisions and sections will tend to improve them; but the men, to learn how to re-form column from square (a movement of the greatest importance before the enemy) will have to practise it in their companies.

There is another point to be considered, and that is the military bearing of the men. We do not only mean the erect, proud, and yet easy position of each individual man under arms, but also that quick simultaneous action in company and battalion movements which is as necessary to a body on the move as to a battalion handling its rifles at a standstill. Volunteers appear quite satisfied if they manage, somehow or other, to get into their proper places in something like the prescribed time, including, generally, a few seconds of respite. No doubt this is the principal point, and in the first year of the existence of a volunteer corps anybody would be perfectly satisfied with it. But there is for every move a certain fixed mode of doing it, prescribed by the regulations, and this is supposed to be that mode by which the object in view can be attained in the shortest possible time, with the greatest convenience to all concerned, and, consequently, with the highest degree of order. The consequence is, that every deviation from the prescribed mode is necessarily connected by a slight degree of disorder and want of regularity, which not only makes an impression of slovenliness upon the beholder, but also implies a certain loss of time, and makes the men think that the detail of the regulations is mere humbug. Let any man see a body of volunteers advance by double files from the centre and front, form company, or go through any other change of formation, and he will at once see what kind of negligent habits we are attaining. But such faults, which may be suffered in an old line regiment, which has a good sub-stratum of solid drill, and will be made to go through the same drill again and shake off its easy ways, are far more dangerous in a body of volunteers, where that solid foundation of detail-drill is unavoidably wanting. Their slovenly habits, which have to be tolerated in the beginning, as the men *must* be hurried through all elementary work, will increase and multiply unless regularly and assiduously checked by strict company drill. It will be impossible to drive such habits out entirely, but at all events they may be, and ought to be, so far checked as not to gain ground. As to the individual bearing of the men, that we suppose will gradually improve, though we very much doubt whether that peculiar waving of a line, marking time, seen in all volunteer drills, will ever disappear. We allude to a certain habit of moving the upper part of the body in marking time, which appears common to all volunteers we have yet seen. No sooner goes up the right foot, than up goes the right shoulder and down goes the left; with the left foot, the left shoulder moves upwards, and thus the whole line waves to and fro like a ripe corn-field under a wild zephyr, but not very much like a body of sturdy soldiers prepared to meet the enemy.

We believe we have said enough to call attention to the subject. Every volunteer who has the movement at heart, will agree with us as to the necessity of regular and diligent company drill; for, let us repeat it, the volunteer force has been unavoidably neglected in its elementary education, and it requires great attention and a deal of work to make up in some manner for this defect.

8. Aldershot and the Volunteers[1]

The Duke of Cambridge, in his speech at the London Rifle Brigade dinner, said he should be very glad to see the volunteers at Aldershot. The only difficulty, to him, appeared to be, how to get them there. We propose to venture a few suggestions how to overcome this difficulty.

It is, undoubtedly, quite out of the question to send to Aldershot or any other camp, whole corps of volunteers. The elements of which they are composed preclude every chance of it. There is no company, much less a battalion, a majority of whose members could spare as much as a fortnight, at one and the same time, for such a purpose.

But if we cannot get the volunteers to Aldershot in bodies, could they not go there singly, and yet learn a great deal? We think they could, if the thing was arranged so as to offer every facility to volunteers to avail themselves of the opportunity.

We believe the great majority of the volunteers to be composed of men who can, now and then, get relieved from their usual avocations for a fortnight in a year. A great many take a regular holiday for that duration, and even longer. Among these there are certainly a considerable number who would not at all object—on the contrary—to spend, for once, their time and their money at Aldershot, if they were received there. Thus, there would be no

[[1] *Volunteer Journal*, Vol. II, No. 36, pp. 93–4, May 11, 1861.]

difficulty whatever, between May and the end of September, to keep at Aldershot a floating population of volunteers amounting, at all times, to the strength of a decent battalion at least. If we can, then, get this floating population to the camp, how can this be utilized?

We propose that a range of huts or tents be set apart for say 600 volunteers, and that a captain, or, better still, a major, from the line be appointed to the command of this volunteer camp, with an adjutant and sergeant-major to assist him. The camp to be opened say in May, as soon as a sufficient number of volunteers have given in their names; if the camp is full, further applicants to be admitted as there may be room for them, the whole of such volunteers to be formed into a battalion; a blouse of a prescribed cut and colour to be worn over the tunics, so as to give the whole a uniform look. As there is sure to be an excess of officers, there will be no other chance but to make officers do duty, for the time being, as sergeants and even privates. Far from considering this a drawback, we should consider it an advantage. No volunteer officer is so well grounded in his personal drill that such a momentary relapse into the ranks would be useless to him; let him recollect that every line officer has to shoulder the rifle for a certain period year after year. The distribution of the temporary officers' posts in the battalion might be easily regulated: the senior captains present might begin, and afterwards others might take their posts by rotation. The major in command could perhaps be intrusted with a deal of discretionary power in nominating to these appointments, in order to ensure a lively emulation among the officers present. These, however, are matters of detail, the arrangement of which would cause but little trouble if the idea was once taken up in good earnest.

Such a battalion, with its floating population, would never attain any very great efficiency, and the major in command, as well as his assistants, would have no easy post of it. But it would ensure one object: that among the volunteer army generally, and among the officers and sergeants specially, a nucleus would be formed of men who have at all events really been soldiers, if only for a fortnight. This may look a contemptibly short period; yet we have no doubt that every man would feel immensely different on leaving, to what he did on reaching Aldershot. There is an immense difference between drilling once or twice a week after the whole day has been spent on business and other matters, and drilling, even for a fortnight only, morning, noon, and night in a camp. During that fortnight, every volunteer present will have no other business to look after but his military education; he will be confirmed in his drill to a degree which no length of the present volunteer drill can raise him up to; and, besides, he will see a great deal more of soldiering than he ever could expect to see in his own corps, unless it encamped on purpose. On leaving Aldershot, every man will think that he has learned during that fortnight at least as much as during the whole of his preceding volunteer service. In due time there will be scarcely a company of volunteers in which one or more members have not been down to Aldershot; and everybody must see to what an extent such an infusion of better-educated elements will improve the steadiness and the military manners, too, of the various corps.

We have supposed that the course of instruction for every man is to be a fortnight, merely because almost everybody might find means to spare that short time. But there could be nothing to prevent allowing such volunteers as can afford it, to stop at the camp for a full month.

As a matter of course, the volunteers in camp would have to keep themselves. The Government ought to find tents and camping utensils, and might, perhaps, make arrangements for the delivery of rations, to be paid for by the men. In this way, without costing the country anything to speak of, the affair would be cheap for the volunteers, and everything put upon a regular camp footing.

We have no doubt that, were the experiment once made, the volunteers would at once cordially respond to it; the battalion would be kept up always to its full complement, and, perhaps, the necessity for similar battalions, at other camps, or at Aldershot, would soon arise. If the excess of officers became very considerable, a special 'officers' battalion' might be formed at one of the camps, with a somewhat longer period of attendance, and we believe such a battalion would answer well for at least one season.

There is, however, another mode of making the camps, and the line generally, useful for volunteer officers: by attaching, temporarily, such officers to battalions of the regulars. This might be done without taking the officers too far from their homes; during the period of such attachment (say a month) the volunteer officer to do duty as if actually serving in the regiment. No doubt means might be found to allow at least one volunteer officer at a time to be thus attached to a battalion, without in any way infringing upon the habits and position of the line officers, who have always shown the best possible spirit towards the volunteers. If this point was taken up, we should consider it advisable to allow no volunteer officer to be attached to the line who had not shown, in some manner or other, that he was fit to profit by it; for he would go there, not to be taught the rudiments, but to be confirmed and perfected in what he knows already, and to learn matters which he cannot learn in his corps.

Both our suggestions—the formation of floating battalions at the camps, and the permission for duly qualified volunteer officers to be attached to the line for a month—have in view the education of the officers chiefly. We repeat, again and again, that the officers form the weak points of the volunteer army; we add, that it now must be evident to all that the present system of volunteer education *cannot make the officers, as a body, efficient*, and that, therefore, new means of instruction must be found if the force is, not only not to retrograde, but to improve.

We throw out these suggestions for no other purpose but to invite attention to the question. We have no wish to lay before the public a fixed plan, with all details worked out, all eventualities provided for, ready to be put into practice at once. That would be the business of others, if the matter was taken up seriously. But we mean to say, the whole volunteer movement was an experiment, and unless people are prepared to experimentalize a little more in order to find out the proper way to improve the new army which has resulted from that experiment, the movement must ultimately come to a deadlock.

9. The War Office and the Volunteers[1]

We believe that in all Great Britain, nowhere has there been among volunteers a greater readiness and alacrity to conform to all War Office orders and regulations, to take up a proper position with regard to the regular army, to work the movement in harmony with the authorities, than in Lancashire, and among other towns, in Manchester. When armouries were ordered to be provided, the

[[1] *Volunteer Journal*, Vol. II, No. 40, p. 125, June 8, 1861.]

order was carried out, although it unavoidably implied great inconvenience in a large town. Whatever orders were sent down were obeyed at once and without a murmur. When our volunteers met in large masses, they anticipated the Duke of Cambridge's desire, and requested the military authorities of the district to take the command and organize the brigades. The desire for efficiency made our Lancashire volunteers criticize all Government interference with a favourable eye; they knew that uniformity and regularity were above all things requisite, and they looked upon every War Office circular as a step towards ensuring these requisites. The *Volunteer Journal*, from its very first number, has not ceased to recommend a willing and cheerful obedience to War Office orders, and to advocate the great advantages of perfect harmony between the volunteers and the military authorities, both local and central. While in other localities, especially in London, there were mysterious rumours abroad respecting the baneful influence of the Horse Guards, the attempts of the authorities to get in the thin end of the wedge, etc., we have never been swayed by such considerations for a single moment. We have given the Commander-in-Chief, the Secretary for War, and all their subordinates, full credit for sincerity when they asserted their willingness to support the movement in every possible shape and form.

But we cannot close our eyes to the fact, that latterly one or two little matters have occurred which look as if there really had been some change in the view men in authority take of the volunteer movement, especially since Lord de Grey and Ripon gave up the Under-Secretaryship for War. A few weeks ago, we believe it was on Whit-Monday, Lord Ranelagh reviewed in Regent's Park such of the London volunteers as would come on his invitation. Now, we have more than once strongly condemned Lord Ranelagh's attempts at playing general. He might have applied to Colonel McMurdo, the inspector-general of volunteers, to review his men, or to recommend another qualified officer for the purpose. However, right or wrong as regards propriety, he went with his men to the park; the affair had been publicly announced, and was so well known that a large crowd of spectators assembled. There were among this crowd people who behaved in a most shameful manner; they pressed round the volunteers, broke their ranks, rendered evolutions impossible, threw stones, and some even, it is stated, attempted to wound the officers' horses with pointed instruments. When this commenced, the officers in charge naturally looked out for the police, but out of the 6,000 men constituting the army of Sir Richard Mayne, we are told *that not one man was there*! The consequence was, that Lord Ranelagh's review was a total failure, owing to the interference of the crowd. Now, if the matter had been allowed to take its course, it is quite possible that it would have proved as much a failure on its own merits, as Lord Ranelagh's previous attempts had invariably done. As it was, Lord Ranelagh was made a martyr of, and strongly recommended to the sympathy of all volunteers.

There can be no doubt that the total absence of the police from this publicly-announced review was not quite accidental. It has been stated in the press that they must have had orders to keep away; and we know that in London, among volunteers, it is very generally believed that the Horse Guards had something to do with this affair, and that it is desired at the Horse Guards to undermine the volunteer movement in every possible way. The feeling in London is very strong upon the matter, and we confess the facts of the case—which, as far as we are ware, it has never been attempted to excuse or explain away—are well adapted to create such a feeling.

This week we have to record another affair which certainly does not look as if the authorities intended to do, as they promised, everything in their power to assist the volunteers. It has been announced, some time ago, that one of our Manchester regiments intended to go into camp for a short period. We believe this announcement was not made before it had been ascertained that it could be carried out. It is commonly reported that verbal application was made to the authorities for tents, etc., and that this application was granted; and that, moreover, the terms had been fixed upon which it had been granted. We believe these arrangements were come to not more than two or three weeks ago. On the strength of this, all other arrangements as to the ground for encampment, canteen, officers' mess, and other matters, were entered into; and when everything is straight, and the formal application for the tents is made, the Government all at once draw back, and declare they cannot furnish any tents at all!

As a matter of course, this upsets the whole plan, and the

expense and trouble incurred by the regiments has all been wasted; and we all know that volunteer regiments have every reason to be careful of their small balance, if any, at the banker's. We are told that so many volunteer regiments are said to have applied for tents that the Government cannot possibly find tents for all, and that therefore none can be furnished to any corps. Whether this be correct or not, the Government ought to know that a bargain is a bargain, and that posterior events could not relieve them from engagements already undertaken. But rumour, which is now beginning to do its work in Manchester quite as much as in London, says that this is a mere idle pretext, and that the Government do not want the volunteers to go under canvas at all; that even if the corps in question were to provide tents or huts at their own expense, and from an independent source, the encampment would not be looked on with a favourable eye in high quarters.

Such occurrences are certainly not adapted to promote that cordiality between the authorities and the volunteers which is so essential to the further success of the movement. The movement is too powerful for any Government to put down; but want of confidence in the authorities on the part of the volunteers, and hidden opposition on the part of the authorities, can very soon create considerable confusion, and hamper its progress for a time. This ought not to be allowed. There are a great many volunteer officers in Parliament. Let them get up in their places and take care that the Government give such explanations as will at once put the matter right, and show the volunteers that they will have to expect cordial support instead of hidden hostility.

10. Volunteer Officers[1]

'Lieutenant A.B., dishonourably discharged; Second Lieutenant C.D., struck off the list; Captain E.F., dismissed the United States service'—such are a few specimens of the latest items of military news we receive by wholesale from America.

The United States have had a very large volunteer army in the field for the last eight months; they have spared neither trouble nor expense to make this army efficient; and, moreover, it has had the advantage of being posted, almost all that time, in sight

[[1] *Volunteer Journal*, Vol. III, No. 64, p. 315, November 22, 1861.]

of the outposts of an enemy who never dared to attack it in a mass or pursue it after a defeat. These favourable circumstances ought to make up, to a very large extent, for the disadvantages under which the United States volunteers were organized; for the poor support they got from a very small army of the line, forming their nucleus; and for the want of experienced adjutants and drill instructors. For we must not forget that in America there were many men both fit and ready to assist in the organization of the volunteers—partly German officers and soldiers who had undergone regular training and seen service in the campaigns of 1848–9, partly English soldiers emigrated during the last ten years.

Now, if under these circumstances a regular weeding of the officers becomes necessary, there must be some weakness inherent, not to the volunteer system in itself, but to the system of officering volunteers by men chosen indiscriminately by themselves from among themselves. It is only after an eight months' campaign in the face of the enemy that the United States Government ventures to call upon volunteer officers to qualify themselves, in some degree, for the duties they undertook to perform when they accepted their commissions; and see what an amount of voluntary or forced resignations, what a heap of dismissals, more or less dishonourable, is the consequence. No doubt, if the United States army of the Potomac were opposed to a force steadied and kept together by a due proportion of professional soldiers, it would have been undoubted individual bravery of the men composing it.

These facts may well serve as a lesson to the volunteers of England. Some of our readers may recollect that, from the very starting of the *Volunteer Journal*, we maintained that the officers were the weak point of the volunteer system, and insisted upon an examination, after a certain time, calling upon the officers to prove that they were at least in a fair way of becoming fit for performing the duties they had undertaken. Most of the gentlemen who had taken upon themselves to command and to instruct men in a line of business of which they were as perfectly ignorant at the time as the men themselves—most of these gentlemen scorned the idea. That was the time when all Government assistance and Government interference were equally scorned. But since then the call upon the pockets of these same gentlemen has been heavy enough to make them apply for pecuniary assistance from Government; and, as Governments run, this means, at the same time, a call for Government interference. Moreover, a two years' experience has brought out pretty plainly the defects of the present system of officering volunteer corps; and we are now informed by a metropolitan commanding officer, and apparently upon authority, that before long the volunteer officers will be called upon to prove their fitness for command, before a board of examination.

We heartily wish this to be the case. The fact is, the English volunteer officers, too, do require weeding to a certain extent. Look at a line regiment at drill and compare it to a volunteer battalion. What it takes the volunteers an hour and a half to go through, the line men go through in less than half an hour. We have seen a deal of square-forming by some of the best volunteer regiments in the country, and we cannot help saying they must be wretched cavalry that would not have cut them up each time before they had their flanks ready for firing. That was not the fault of the men. They appeared to know their duty as well as could be expected, and to do it sometimes even as mechanically as you see in a line regiment. But the men had to wait for the company officers, who appeared to hesitate about the word of command to be given, and about the moment when they ought to give it. Thus, hesitation and sometimes confusion was thrown into a formation which, above all others, requires a promptness, both of command of execution, imparted by long practice only. Now, if this be the case after two years' practice, is this not a proof that there are plenty of volunteer officers holding responsible situations which they are not fit to hold?

Again, the commanders of battalions have lately received some very high praise from the hands of highly competent authority. It was said that they appeared to be up to their work, while the company officers were not always so. We are not at all inclined, as will have been seen above, to dispute the latter statement; but we must say that if the high authority alluded to had seen the lieutenant-colonels and majors, not at a great review, but at plain battalion drill, the opinion given would probably have been slightly different. At a great review, no field officer in command of a battalion, if not perfectly up to his work, would attempt to act on his own responsibility. He has his adjutant—who knows what he is about—for a prompter; and he is prompted by him accordingly, and goes through his work creditably, while the poor captain has to bungle through his performance without any prompter at all. But look at the same field officer at battalion drill. There he has no vigilant general's eye watching him; there he reigns supreme; and there the adjutant, often enough, has to take the post assigned to him by the Queen's regulations, and must keep his advice to himself until asked for it, or until the mess is complete. This is the place where you see the volunteer field officer in his true light. He is there to instruct his men in battalion drill; but not being himself perfect in that science, he profits of their being there to instruct himself in it. As the old saying goes, *docendo discimus*. But if the teacher is not well on his legs in the art he has to teach, blunders and confusion are apt to occur, and, unfortunately, do occur often enough. It will not contribute either to the proficiency in drill of a volunteer battalion, or to its confidence in its commander, if the men find out that battalion drill, for them, means nothing but giving their field officer in command an opportunity of learning his drill himself, while they are tossed about here and there, without any purpose even, and expected to rectify, by their superior knowledge, the blunders of their superior officer.

We do not mean to say that commanding officers of volunteers have not put themselves to some trouble to learn their duty; but we do mean to say that if company officers cannot be manufactured out of civilians as easily as private soldiers, field officers are far more difficult to manufacture. We must come to the conclusion, on the mere ground of battalion drill experience, that none but professional soldiers are fit to command battalions. And if we consider that drill is but one part of a field officer's duty, that the commander of a battalion, being liable to be detached for independent duty, where he has to act on his own responsibility, requires a knowledge of higher tactics, we must say that we should be very sorry to see the lives of 600 or 1,000 men entrusted to the guidance of such civilians as now form the great majority of commanders of battalions.

Depend upon it, if the English volunteers ever will have to face an enemy, it will not be under the favourable circumstances which now permit the American Government to clear the ranks of their volunteer officers from the most incapable subjects. If the English volunteers are called out, it will be to fight, not a volunteer army like themselves, but the most highly disciplined and most active army in Europe. The very first engagements will be decisive; and, depend upon it, if any hesitation or confusion arises, either by the wrong commands of the colonels, or by the colonels, or by the uncertainty of the captains, that will be taken advantage of at once. There will be no time for weeding when once before the enemy, and therefore we hope it will be done while there is time.

PART II
THE HISTORY OF THE RIFLE

11. The History of the Rifle—I[1]

The rifle is a German invention, dating as far back as the close of the fifteenth century. The first rifles were made with apparently no other object than to facilitate the loading of the arm with an almost tight-fitting bullet. To this end, the grooves were made straight, without any spiral turning, and merely served to diminish the friction of the bullet in the bore. The bullet itself was surrounded by a piece of greased woollen or linen cloth (the plaster), and was thus hammered down without too much difficulty. These rifles, primitive as they were, must have given far better results than the smooth-bore small arms of the period, with their bullets of considerably smaller diameter than the bore.

Later on, the character of the arm was totally altered by the spiral turn given to the grooves, which transformed the bore of the barrel into a sort of a female screw; the bullet, by the tight-fitting plaster, being made to follow the grooves, took the spiral turn as well, and thus retained a spiral rotation round its line of flight. It was soon found that this mode of fixing the rotation of the bullet vastly increased both the range and accuracy of the arm, and thus the spiral grooves very soon superseded the straight ones.

This, then, was the kind of rifle which remained in general use for more than two hundred years. If we except hair-triggers and more carefully worked sights, it scarcely underwent any improvement up to 1828. It was greatly superior to the smooth-bore musket in accuracy, but not so very much in range; beyond 300 or 500 yards, it could not be relied upon. At the same time, it was comparatively difficult to load; the hammering down of the bullet was a very tedious operation; the powder and plastered bullet

[[1] *Volunteer Journal*, Vol. I, No. 9, pp. 123–4, November 3, 1860. Article by the author of 'A German Account of the Newton Review'.]

had each to be put separately into the barrel, and not more than one round per minute could be fired. These drawbacks made it unfit for the greater generality of an army, especially at a time like the eighteenth century, when all battles were decided by the rapid firing of deployed lines. With such tactics, the old smooth-bore musket, with all its glaring imperfections, was still a far preferable arm. Thus we find that the rifle remained the favourite implement of the deer-stalker and chamois-hunter, and that it was used as an exceptional arm of war, for a few battalions of sharp-shooters, in such armies only as could recruit these battalions from a sufficient number of trained sportsmen among the population.

The wars of the American and French Revolutions created a great change in tactics. Henceforth extended order was introduced in every engagement; the combination of skirmishers with lines or columns became the essential characteristic of modern fighting. The masses, during the greater part of the day, are kept back; they are held in reserve or employed in manœuvring so as to concentrate on the weak point of the enemy; they are only launched in decisive moments; but, in the meantime, skirmishers and their immediate supports are constantly engaged. The mass of the ammunition is spent by them, and the objects they fire at, are seldom larger than the front of a company; in most cases, they have to fire at single men well hidden by covering objects. And yet, the effect of their fire is most important; for every attack is both prepared, and, in the first instance, met by it; they are expected to weaken the resistance of detachments occupying farm houses or villages, as well as to take the edge off the attack of a charging line. Now, with old 'Brown Bess', none of these things could be done effectively. Nobody can ever have been under the fire of skirmishers armed with smooth-bore muskets, without taking home an utter contempt for its efficiency at medium ranges. Still, the rifle in its old shape was not fitted for the mass of skirmishers. The old rifle, in order to facilitate the forcing down of the bullet, must be short, so short that it was but a poor handle to a bayonet; consequently, riflemen were used in such positions only when they were safe against an attack with the bayonet, or by cavalry.

Under these circumstances, the problem at once presented itself: to invent a gun, which combines the range and accuracy of the rifle, with the rapidity and ease of loading, and with the length of barrel of the smooth-bore musket; an arm, which is at the same time a rifle and a handy arm of war, fit to be placed into the hands of every infantry soldier.

Thus we see that with the very introduction of skirmishing into modern tactics, arose the demand for such an improved arm of war. In the nineteenth century, whenever a demand for a thing arises, and that demand be justified by the circumstances of the case, it is sure to be supplied. It was supplied in this case. Almost all improvements in small arms made since 1828 tended to supply it.

Before, however, we attempt to give an account of those improvements which have created such great and numerous changes in rifled fire-arms, by dropping the old system of forcing the bullet home, we may be allowed to cast a glance at the attempts made to improve the rifle while maintaining the old method of loading.

The rifle with oval bore which is known in England as the Lancaster rifle, has been in use on the Continent for more than forty years. We find it mentioned in a German military book printed in 1818. In Brunswick, Colonel Berger improved it and had the whole infantry of that duchy armed with it in 1832. The ovality was but slight, and the oval bullet was forced home in the old fashion. This oval bullet, however, was to be used in skirmishing only. For volley firing, the men were provided with spherical bullets of smaller calibre, which rolled down the barrel quite as easy as any musket ball. Still, the inconveniences of this system are obvious. It is merely remarkable as the first instance of giving rifled muskets to the whole of the infantry in any one army.

In Switzerland, a civil engineer and officer of rifles, M. Wild, improved the rifle considerably. His bullet was smaller in proportion to the bore than usual, and was made to take the rifling by means of the plaster only; a disc on the ramrod prevented it from entering the bore too deep, and thus driving the bullet so close on the charge that the powder got crushed; the spirality of the grooves was reduced and the charge increased. Wild's rifle gave very good results up to above 500 yards, with a very flat trajectory; besides, it allowed of more than 100 shots being fired without fouling. It was adopted in Switzerland, Württemberg, and Baden, but is now, of course, antiquated and superseded.

The most modern and the best rifle constructed upon the forcing principle is the new Swiss sharpshooter's regulation rifle. This arm has adopted the American principle of a very small calibre; its bore is not more than 10·50 millimetres, or 0·42 of an inch. The barrel is but 28 inches long, and has eight flat grooves (one turn in 34 inches). The ramrod is provided with the disc as introduced by Wild. The bullet is cylindro-ogival, and very long; it is forced home by means of a greased plaster. The charge is comparatively strong, and of a very coarse-grained powder. This arm has shown the most astonishing effects; and in the trial of various rifles recently made by the Dutch government, its range, accuracy and lowness of trajectory were found to be unequalled. In fact, at a range of 600 yards, the highest point of its trajectory is only 8 feet 6 inches, so that the whole of the flight, at that range, is dangerous space for cavalry, and that even for infantry the last 100 yards of the trajectory are dangerous space; in other words, an error in judging distance of 100 yards, at 600 yards range, would not prevent the bullet from hitting an object six feet high. This is a result far surpassing that of any other rifled musket; the very best of them require an elevation, which raises the highest points of the trajectory, for 600 yards, to 13 to 20 feet, and reduces the dangerous space from 60 to 25 yards. This extraordinary flatness of trajectory is produced by the small calibre of the arm, which admits of a very elongated bolt-shaped shot, and of a comparatively powerful charge; with a small bore, the rifle may be made very strong, without being clumsy, the shot may be long, without being heavy, and the charge may be powerful, relatively, without producing too severe a recoil. It is certain that the forced loading has nothing to do with the admirable shooting of the arm; indeed, it forms its only drawback, and prevents it from being used as the general arm of infantry. The Swiss have, therefore, restrained it to their companies of sharpshooters, in whose hands, no doubt, it will answer uncommonly well.

In the next article, we shall show how the rifle came to be made into a weapon fit to be placed into the hands of every infantry soldier.

12. The History of the Rifle[1]—II

Delvigne, a French officer, was the originator of the first attempt to make the rifle a weapon fit for general infantry use. He saw clearly that to do this, the bullet must slip down the barrel as eas[il]y, or nearly so, as the bullet of a smooth-bore musket, and be made, afterwards, to change its shape so as to enter into the grooves.

To obtain this end, he constructed, as early as 1828, a rifle with a chamber at the breech; that is to say the extreme end of the bore at the breech, where the powder lies, was made of considerably smaller diameter than the remaining part of the barrel. This chamber was adopted from howitzers and mortars which had always been so constructed; but while, in ordnance, it merely served to keep well together the small charges used for mortars and howitzers, it answered quite a different purpose in Delvigne's rifle. The powder having been dropped down into the chamber, the bullet, smaller than the bore, was made to roll down after it; but, arrived on the edge of the chamber, it could not pass any further, and remained supported on it; and a few smart blows with the ramrod were sufficient to force the soft lead of the bullet sideways into the grooves, and to enlarge its diameter so much that it fitted tight in the barrel.

The greatest inconvenience in this system was, that the bullet lost its spherical shape, and became somewhat flattened, in consequence of which it was apt to lose the lateral rotation impressed upon it by the grooves, which impaired its precision considerably. To remedy this, Delvigne invented elongated shot (cylindro-conical), and although the experiments with this kind of shot were not, at first, very successful in France, it answered very well in Belgium, Austria, and Sardinia, in which countries Delvigne's rifle, with various improvements, was given to the Chasseur battalions instead of the old rifle. Although his rifle is at present almost everywhere superseded, Delvigne's improvements embrace the two great principles on which all succeeding inventors have been obliged to rely. Firstly, that in muzzle-loading rifles, the shot must go down with a certain windage, so as to admit of easy loading, and must change its shape, so as to enter the grooves,

[1 *Volunteer Journal*, Vol. I, No. 11, pp. 153–4, November 17, 1860. Article by the author of 'A German Account of the Newton Review'.]

only after it has been rammed home; and secondly, that elongated shot are the only projectiles adapted for modern rifles. Delvigne thus at once put the question on its proper footing, and fully deserves the name of the father of the modern rifle.

The advantages of elongated shot over spherical bullets are numerous, so long as their lateral rotation (around its longitudinal axis) can be secured to the former, which is accomplished in a satisfactory manner by almost every modern rifle. The elongated shot offers a far smaller section, in proportion to its weight, to the resistance of the atmosphere than the spherical bullet. Its point can be so shaped as to reduce that resistance to a minimum. Like a bolt or an arrow, it is to a certain degree supported by the air below it. The consequence is, that it loses far less of its initial velocity by the resistance of the air, and that, consequently, it will reach a given distance with a far lower trajectory (that is to say, with a line of flight far more dangerous to the enemy) than any round shot of the same diameter.

Another advantage is, that the elongated shot offers a far greater surface of contact to the sides of the barrel than the round shot. This makes the former take the rifling far better, and therefore admits of a reduced pitch of the rifling as well as of a reduced depth of groove. Both these circumstances facilitate the cleaning of the arm, and at the same time permit the use of full charges without increasing the recoil of the gun.

And finally, as the weight of the elongated shot is so much greater than that of the round bullet, it follows that the calibre, or diameter of bore, of the gun can be considerably reduced, while it still remains capable of firing a projectile equal in weight to the old round bullet. Now, if the weight of the old smooth-bore musket and that of its bullet be considered as the standard weights, a rifle for elongated shot of this weight can be made stronger than the old musket in the same proportion as the bore has been reduced, and it will still not exceed the weight of the old musket. The gun being stronger, it will stand the charge so much the better; it will have less recoil, and, consequently, the reduced bore will admit of relatively stronger charges, whereby a greater initial velocity, and, consequently, a lower line of flight will be secured.

The next improvement was made by another French officer, Colonel Thouvenin. He clearly perceived the inconvenience of leaving the shot, while being rammed into the grooves, supported on a circular projection touching its edges. He therefore did away with the edges of the chamber, boring out the whole of the bore to one uniform diameter as heretofore. In the middle of the screw closing the bore, he fixed a short strong steel pin, or peg, which projected into the bore, and around which the powder was to fall; on the blunt top of this peg the shot was to be supported while the ramrod hammered it into the grooves. The advantages of this system were considerable. The expansion of the shot, by the blows of the ramrod, was far more regular than in Delvigne's rifles. The arm could afford a greater windage, which facilitated loading. The results obtained with it were so satisfactory that, as early as 1846, the French *Chasseurs à pied* were armed with Thouvenin's rifles; the Zouaves and other light African infantry followed; and as it was found that the old smooth-bore muskets could, with little expense, be transformed into Thouvenin's rifles, the carbines of the French foot-artillery were all altered accordingly. The Prussian rifles were armed with Thouvenin's rifle in 1847; those of Bavaria in 1848; and most of the smaller States of Northern Germany followed the example, in some cases arming even portions of the line with this excellent weapon. In all these rifles there is visible a certain approach to unity of system, in spite of all their variations as to calibre, etc.; the number of grooves is reduced (mostly to 4), and the pitch generally is from three-quarters of a turn to one turn in the whole length of the barrel.

Still, Thouvenin's rifle had its drawbacks. The force required to drive, by repeated blows, the lead of the shot, laterally, into the grooves was incompatible with that length of barrel which the common musket of infantry of the line must always have as an effective handle to a bayonet. It was, besides, very difficult for skirmishers, crawling or kneeling, to apply that force. The resistance offered to the explosive force by the shot, jammed as it is in the grooves just in front of the powder, increases the recoil, and thereby restricts the gun to a comparatively small charge. Finally, the peg always remains an undesirable complication of the arm; it rendered the cleaning of the space around it very difficult and is liable to get out of order.

Thus the principle of compressing the shot by blows from the ramrod gave very satisfactory results, for the time being, in the system of Delvigne, and better results, again, in that of Thouvenin.

Still it could not assert its superiority, for an arm for general infantry use, over the old smooth-bore gun. Other principles had to be resorted to before a rifle fit for every soldier's hands could be produced; and of them we shall speak in a succeeding paper.

13. The History of the Rifle[1]—III

Delvigne, whose rifle we described in the preceding article, found it advisable to hollow out his elongated bullets from the base, in order to reduce their weight to something like that of the old spherical bullet. Though he very soon found that this hollow projectile was incompatible with the system of expanding the shot by mechanical blows, his experiments sufficed to prove to him that the gas developed by the explosion, on entering the hollow formed in the bullet, had a tendency to expand the walls of this hollow portion so as to make the bullet fit the barrel exactly and take the rifling.

It was this discovery which was taken up in 1849 by the then Captain Minié. He did entirely away with the peg or pillar at the bottom of the bore, and restored to the rifle the simplicity which it had possessed before Delvigne and Thouvenin; relying entirely upon the expansive action of the explosion upon the hollow portion of his bullet. This bullet was cylindro-ogival, with two ring-shaped indentations round the cylindrical portion,[2] and hollowed out conically from the base; a cup-shaped hollow iron plug (*culot*) closed the hollow portion, and was driven into it by the force of the explosion, thereby effectively expanding the lead. The bullet had sufficient windage to go easily down, even when surrounded by the greased paper cartridge.

Here, then, we have at last a rifle and a bullet constructed upon principles which render it possible to give this arm to every foot soldier. The new arm loads as easily as the smooth-bore musket, and has an effect far superior to that of the old rifle, which it equals in precision, but far exceeds in range. The rifle with expansion-bullet is undoubtedly—of all muzzle-loaders—the best arm for

[[1] *Volunteer Journal*, Vol. I, No. 14, p. 188, December 8, 1860. Article by the author of 'A German account of the Newton Review'.]

[2] These indentations (*cannelures*) had been invented by Tamisier, another French officer. Besides reducing the weight of the bullet and the friction in the barrel, they were found to balance the shot in the air, similar to the wings of an arrow, and thus to lower the trajectory.

general use as well as for sharpshooters, and it is owing to this circumstance that it owes its very great success, its adoption in so many services, and the many attempts that have been made to improve the shape of the shot or the grooving of the rifle. The Minié bullet, in consequence of its being hollowed out, can be made but little heavier than the old round bullet of the same calibre; the bullet lying loose on the powder, and being only gradually expanded as it passes through the barrel, the recoil is far less than with either the old or the Delvigne and Thouvenin rifles, in every one of which the shot is jammed fast in the barrel, and has to be dislodged by the full force of the explosion; consequently the Minié rifle can apply a relatively stronger charge. The grooves have to be made very shallow, which facilitates the cleaning of the barrel; the length of axis in which one full turn of the grooves is made has to be pretty great, in consequence of which the number of rotations, and also the friction with the air (which takes place at every rotation), is diminished, whereby the initial velocity is better preserved. The hollow base-end of the shot also brings its centre of gravity more forward; and all these circumstances combine to produce a comparatively low trajectory. The general adoption of the Minié rifle was, in fact, owing to another circumstance: that, by a very simple process, all old smooth-bore muskets could be transformed into rifles fit for Minié bullets. When the Crimean war made it desirable, in Prussia, that the whole infantry should at once be armed with rifled muskets, and the requisite number of needle-guns had not yet been manufactured, 300,000 old muskets were rifled and rendered fit for Minié ammunition in less than a year.

The French Government were the first to arm a few battalions with the Minié rifles; but the grooves were progressive, that is to say they were deeper at the breech than at the muzzle, so that whatever lead had entered the grooves at the breech, was again compressed by the shallowing groove during its progress through the barrel, while at the same time from within the expanding force of the powder continued to act. Thus such an amount of friction was created that very often the solid point of the shot was torn off and sent out of the barrel while the hollow base-end remained fast in the grooves. This, and other defects, induced the Government to renounce any further attempt to introduce the Minié rifle.

In England, as early as 1851, 28,000 of these rifles were constructed, similar to those tried in France; the bullet was slightly conical, with ogival point, with a round hollow plug, and without indentations, as it was intended to press them. The results were very unsatisfactory, chiefly in consequence of the shape of the bullet; until, in 1852, new experiments were made, from which, finally, the Enfield rifle and bullets proceeded, which will be again alluded to hereafter. The Enfield rifle is but one of the modifications of the Minié. It has, since 1854, definitely superseded all smooth-bore muskets in the British army.

In Belgium, the Minié rifle, with slight alterations, has been adopted since 1854 for riflemen, and latterly for the line also.

In Spain, in 1853, the rifles received the Minié, which has since also been given to the line.

In Prussia, in 1855–6, the Minié rifle was provisionally given to the line, as stated above. It has since been completely superseded by the needle-gun.

In the smaller German States, the Minié rifle was also adopted, with very few exceptions.

In Switzerland, the Prélat rifle, destined to arm the whole of the infantry with the exception of the sharpshooters, is but a modification of the Minié.

And in Russia, finally, the Government is just now occupied in replacing the old smooth-bore muskets by Minié rifles of a very good model.

In almost every one of these countries has the number, depth, and pitch of the grooves, and the shape of the bullet, undergone various modifications of detail, to describe the most important of which will be the purport of our next article.

14. The History of the Rifle[1]—IV

We again recapitulate the principle of Minié's system: A rifled musket, with shallow grooves, is loaded with an elongated bullet, which is so much smaller in diameter than the bore, that it glides down easily. This bullet is hollowed out from its base, that is to say, from the end resting on the powder. On firing, the gas suddenly developed by the explosion enters into this hollow part,

[[1] *Volunteer Journal*, Vol. I, No. 15, pp. 199–200, December 15, 1860. Article by the author of 'A German Account of the Newton Review'.]

and by its pressure against its comparatively thin sides, expands the lead so as to make it fit the bore and enter into the grooves, and retain the lateral rotation characteristic of all rifle bullets. This is the principle, the essential part in all the different rifles firing expansion bullets; and it is common to them all. But in matters of detail, a great many modifications have been made by various inventors.

Minié himself adopted the plug. This plug was a little round, cup-shaped piece of sheet-iron, driven into the mouth of the hollow part of the bullet. It was intended to be driven deeper into the hollow by the explosion, and thus to assist and render more certain the expansion of the shot. It was, however, soon found that this cup-shaped plug had great inconveniences. It separated very often from the bullet on leaving the muzzle, and in its irregular line of flight it slightly wounded sometimes troops belonging to the firing party and placed a little in advance laterally. It also sometimes turned over while being driven into the lead, and thus caused an irregular expansion, and thereby a deviation of the shot from the line of aim. As it had been proved that the expansion of the shot might be obtained without any plugs at all, experiments were made to fix the best shape of an expansion bullet without plug. The Prussian Captain Neindorff appears to have been the first to propose such a bullet (in 1852). The hollow of this projectile is cylindrical, but widened out towards the base, in the shape of a tun-dish. This shot gave very good results as to range and precision, but it was soon found that the plug served another purpose besides expansion, it preserved the thin sides of the hollow shot from getting crushed during transport and rough handling; while Neindorff's bullets became deformed during transport, and then gave very bad results. In most German services, therefore, the hollow iron plug was maintained, but it was made of a long, pointed, sugar-loaf shape, and then answered very well, never turned over, and scarcely ever got separated from the leaden shot. The Enfield bullet, as is well known, has a solid wooden plug.

In some States, however, the experiments with bullets without plugs were continued, and such bullets adopted for the service. This was the case in Belgium, France, Switzerland, and Bavaria. The chief object in all these experiments was to fix a shape for the hollow part of the bullet which would prevent crushing while it allowed expansion. Thus the hollow was formed in the shape of a bell (Timmerhans in Belgium), of a three-sided prism (Nessler in France), with a cross-shaped section (Ploennies in Darmstadt), etc. But it appears almost impossible to unite the two elements, solidity and expansibility, in any modification of an expansion shot without a plug. The new Bavarian projectile (Major Podewils'), which has a plain cylindrical hollow, and very strong sides to it, appears, so far, to answer best.

In countries where old smooth-bore muskets were rifled for Minié bullets, the large calibre of the old musket became, of course, compulsory. But where entirely new rifles were provided for the army, the calibre was considerably reduced, from considerations to which we have alluded in a former article. The English Enfield rifle has a calibre of 14·68 millimetres, the South German rifle (adopted in Württemberg, Bavaria, Baden, and Hesse-Darmstadt) of 13·9 mm. The French alone, in their rifles for the guard, retained the calibre of their smooth-bore muskets (17·80 mm.).

The Enfield rifle is a very fair specimen of the expansion system. Its calibre is small enough to admit of a shot twice the length of its diameter, and still not heavier than the old round musket bullet. Its workmanship is very good, and superior to that of almost all rifles served out to Continental troops. The bullet has very good proportions; against the wooden plug it is objected that it may either swell and thereby increase the diameter of the shot, or shrink, and then fall out; but we think these objections futile. If the swelling of the plug presented any inconvenience, it would have been found out long since; and in case of its shrinking, the make of the cartridge prevents its falling out. The results obtained with the Enfield rifle are about on a par with those of the best Continental expansion rifles.

The objections to the Enfield, as a rifle with expansion bullets, are these: that the calibre might still be smaller, giving a longer bullet for the same weight and a stronger barrel with the same weight; that five grooves are proved to be better than three; that the barrel of the long Enfield, at least, is too delicate, towards the muzzle, to be used as a handle for a bayonet; that the bullet, from having no ring-shaped indentations, must suffer an enormous amount of friction in the barrel, and thereby runs the risk of having the solid point torn off, while the ring-shaped hollow part sticks fast to the grooves,

To change the calibre is a very serious matter; and without that it will be very difficult to give the muzzle end of the barrel more solidity. This appears to us the most serious objection. All other objections appear unimportant; the number of grooves and the shape of the bullet may be altered any time without inconvenience; and even as it is, the Enfield has proved itself a very useful arm of war.

We have, so far, compared the Enfield with such rifles only which use expansion bullets; the comparison with rifles based upon different principles we must reserve for a future occasion, when we shall have examined the various other constructions now in use.

15. The History of the Rifle[1]—V

In 1852, an English gun-maker, Mr. Wilkinson, and an Austrian officer of artillery, Captain Lorenz, simultaneously, but each independently of the other, invented another method of making a loose-fitting elongated bullet increase its diameter by the force of the explosion, so as to make it fit the bore closely and follow the turn of the grooves. This method consisted in making the explosion compress the bullet lengthways, instead of expanding it.

Take a soft or elastic ball, place it on a table, and make it fly off with a smart blow of the hand. The first effect of the blow, even before it starts the ball, will be a change in its shape. Light as it is, the weight of the ball offers resistance enough to become flattened on the side where it receives the blow; it is compressed in one direction, and, consequently, its size must increase in another direction, similar to what it does when you completely flatten it. As the blow acts upon the elastic ball, so is the explosion of the powder expected to act upon the compression bullet of Lorenz and Wilkinson. The weight, the *vis inertiae* of the bullet is made the means which, by its resistance to the force of the explosion, compresses the bullet lengthways, and thereby makes it larger sideways; when the shot comes out, it is shorter and thicker than when it was put in.

An elongated bullet of solid lead, in order to offer sufficient resistance, and thus to be sufficiently compressed to take the

[[1] *Volunteer Journal*, Vol. I, No. 17, pp. 223–4, December 29, 1860. By the author of 'A German Account of the Newton Review'.]

grooves, would have to be very heavy—in other words, very long in proportion to its thickness. Even with a small calibre such a bullet would be too heavy for war, as the men would be overweighted with ammunition if they carried the usual number of rounds. To remedy this, two very deep ring-shaped indentations are cut into the cylindrical part of the bullet. Take an Enfield bullet, remove the plug, fill the cavity with molten lead, and when cold, cut these two indentations, close to each other and close to the flat end, into the cylindrical part of the projectile, leaving the three remaining portions of the bullet strung, as it were, upon a common axis of solid lead. The bullet will then consist of two very flat truncated cones, pointing forward, and of the heavy solid point, all of which are solidly connected with each other. This bullet will answer as a compression bullet. The resistance against the explosion will be offered by the heavy fore part or point of the bullet; the head of the rear cone will be driven, by the force of the powder, into the base of the cone in front of it, whose head, again, will be driven into the rear end of the point; and thus the shot, being shortened and compressed in the direction of its length, will be made so much thicker that it closes on all sides to the bore and takes the rifling.

From this it is evident that the solid point is the principal portion of the compression bullet. The longer and heavier it is, the more resistance will it offer, and, consequently, the more certain will be the compressive effect of the explosion. So long as the calibre of the rifle is small, say rather less than the Enfield, it will be possible to make compression bullets not heavier in metal than expansion bullets; but with the calibre grows the surface of the base of the bullet, or in other words the surface exposed to the immediate action of the powder; and this is the cause why, with large calibres, compression bullets will always have to be too heavy to be of any use; otherwise the force of the explosion, by overcoming the resistance of the bullet, would throw it out of the barrel before it had time to become properly compressed. Large-calibred, smooth-bore muskets may, therefore, be altered into rifles for expansion shot, but they will never do for compression bullets.

With small calibres and flat grooves, the compression system gives excellent results. The forward position of the centre of gravity is very favourable to a low trajectory. The compression bullet has all the advantages of the expansion system, as far as regards case and rapidity of loading, and smallness of recoil. The bullet is solid, and can stand transport and rough usage well enough; its shape allows of its being pressed, instead of cast. The only drawback is that it requires a very small windage, of not more than about 0·01 of an inch, and a great regularity both in the size of the bores and that of the bullets, as evidently the compressive effect does not increase the circumference of the shot by near as much as the expansive effect; and thus, with a greater windage, or old barrels, it would be doubtful whether the bullet becomes compressed enough to take the rifling. But this small windage is no great objection, as many rifles with expansion shot have no greater windage (the Enfield, too, for instance, has only 0·01 of an inch), and there is now no difficulty in constructing both barrels and bullets of very exact and regular dimensions.

The Austrian army has adopted the compression bullet for the whole of the infantry. The calibre is small, 13·9 millimetres, or 0·546 of an inch (0·031 less than the Enfield); the barrel has four very flat grooves (an even number of grooves, though decidedly objectionable in expansion rifles, is found to answer better in compression rifles than an odd number), with one turn in about six feet six inches (almost the same as the Enfield). The bullet weighs about 480 grains (50 grains less than the Enfield), and the charge is $\frac{1}{6}$th of its weight (with the Enfield, about $\frac{1}{8}$th of the weight of the bullet). This arm stood its trial in the Italian campaign of 1859, and the great number of French soldiers, and especially officers, who succumbed to it, testify to its excellence. It has a considerably lower trajectory than the Enfield, which is owing to the proportionally stronger charge, to the smaller calibre producing a more elongated shot, and, maybe, to the action of the two ring-shaped indentations.

Saxony, Hanover, and one or two small German States have also adopted, for their light infantry, rifles from which compression bullets constructed on Lorenz's principle are fired.

In Switzerland, besides the sharpshooter's rifle mentioned before, there as been adopted a rifle of the same calibre (10·51 millimetres or 0·413 of an inch, 0·164 smaller than the Enfield) for compression shot. This rifle is used by the light companies of the infantry battalions. The bullet is on Lorenz's model, and the results given by this rifle, in lowness of trajectory, range and precision, class it second only to the Swiss sharpshooter's rifle above alluded to, whose bullet, forced home in the old fashion, has the flattest trajectory of any known rifle. At 500 yards, the Swiss compression bullet fired from this rifle gives a dangerous space of 130 yards![1]

So far, there can be no doubt that the compression-system has given better results than the expansion system, as it has hitherto certainly produced the lowest trajectory. It is, however, equally doubtless that this is not owing to the system in itself, but to other causes, among which the smallness of the calibre is the principal one. With an equally small calibre, the expansion bullet must produce as low a line of flight as its hitherto more successful competitor. This will soon be made evident. The rifles of the four States of South-Western Germany (Bavaria, etc.) have the same calibre as those of Austria, so that they may in case of need use Austrian ammunition, and *vice versa*. But, in adopting the same diameter of bore, they have all of them adopted expansion bullets; and the practice tables of both classes of shot will thus afford a fair test of the merits of either. If, as we expect, the expansion bullet will then give as good results as its competitor, it will deserve the preference; for—1st, it is more certain of taking the rifling, under any circumstances; 2nd, it may be made lighter, with the same bore, than compression shot; and, 3rd, it is less affected by the enlargement of the bore, which takes place in all gun-barrels after having been in use for a certain time.

16. The History of the Rifle[2]—VI

All the rifles which we have hitherto described were muzzle-loaders. There have been, however, in former times, a great many kinds of fire-arms which were loaded at the breech. Breech-load-

[1] By dangerous space is understood that portion of the flight of a bullet in which it is never higher than the height of a man, say six feet. Thus, in this instance, a shot aimed at the bottom of a target six feet high and 500 yards distant, would hit any object, six feet high, standing in the line of aim anywhere between 370 and 500 yards from the man firing. In other words, with the 500 yards sight, an error of 130 yards in judging the distance of the object may be made, and still the object will be hit if the line of aim was taken correctly.

[[2] *Volunteer Journal*, Vol. I, No. 18, pp. 235–6, January 5, 1861. Article by the author of 'A German Account of the Newton Review'.]

ing in cannon preceded muzzle-loading; and most old armouries will contain rifles and pistols two or three hundred years old, with a movable breech, into which the charge could be introduced without being passed through the barrel by a ramrod. The great difficulty always was to join the movable breech in such a way to the barrel that it could be easily separated and put on again, and that the mode of fixing it was solid enough to stand the explosion. With the deficient mechanical contrivances of those times, it was no wonder that these two requisites could not be combined. Either the parts fixing the breech on to the barrel were deficient in solidity and durability, or the process of unfixing and re-fixing it was fearfully slow. No wonder, then, that these arms were thrown aside, that muzzle-loading did its work quicker, and that the ramrod ruled supreme.

When, in modern times, military men and gun-makers were bent upon the construction of a fire-arm which should combine the quick and easy loading of the old musket with the range and precision of the rifle, it was natural that breech-loading should again receive attention. With a proper system of fixing the breech, all difficulties were overcome. The shot, a little larger in diameter than the bore, could then be placed, together with the charge, in the breech, and on being pushed forward by the explosion, would press itself through the bore, fill the grooves with its excess of lead, take the rifling, and exclude all possibility of windage. The only difficulty was the mode of fixing the breech. But what was impossible in the 16th and 17th centuries need not be despaired of now.

The great advantages of a breech-loader, supposing that difficulty overcome, are obvious. The time required for loading is considerably reduced. No drawing, turning round, and returning ramrod. One motion opens the breech, another brings the cartridge into its place, a third closes the breech again. A rapid fire of skirmishers, or a quick succession of volleys, so important in many decisive circumstances, are thus secured in a degree which no muzzle-loader can ever equal. With all muzzle-loaders the art of loading is rendered difficult as soon as the soldier, in skirmishing, is kneeling or laid down behind some covering object. If he keeps behind his shelter he cannot hold his gun in a vertical position, and a great part of his charge will stick on to the sides of the bore while running down; if he holds his gun straight up he has to expose himself. With a breech-loader he can load in any position, even without turning his eye from the enemy, as he can load without looking at his gun. In line, he can load while advancing; pour in volley after volley during the advance, and still arrive upon the enemy with a gun always loaded. The bullet can be of the simplest construction, perfectly solid, and will never run any of the chances by which both compression and expansion shots miss taking the grooves, or experience other unpleasant accidents.

The cleaning of the gun is uncommonly facilitated. The chamber, or place where the powder and bullet lie, which is the part always most exposed to fouling, is here laid competely open, and the barrel or tube, open at both ends, can be easily inspected and cleaned to perfection. The parts about the breech being necessarily very heavy, as otherwise they could not withstand the explosion, bring the centre of gravity of the rifle nearer the shoulder, and thereby facilitate a steady aim.

We have seen that the only difficulty consisted in the proper closing of the breech. There can be no doubt that this difficulty has now been fully overcome. The number of breech-loaders brought out during the last twenty years is wonderful, and some of them, at least, fulfil all reasonable expectations, both as to the efficiency and solidity of the breech-loading apparatus, and as to the ease and rapidity with which the breech can be fixed and unfixed. As arms of war, however, there are at present only three different systems in use.

The first is the gun now used by the infantry in Sweden and Norway. The breech-loading apparatus appears to be sufficiently handy and solid. The charge is fired by a percussion cap, both cock and piston being at the under side of the chamber piece. Of the practice made by this gun we have not been able to obtain any particulars.

The second is the revolver. The revolver, same as the rifle, is a very old German invention. Centuries ago, pistols with several barrels were made, provided with a revolving apparatus, which, after every shot, made a fresh barrel turn into the position required for the action of the lock upon it. Colonel Colt, in America, again took up the idea. He separated the chambers from the barrels, so that one barrel did for all the revolving chambers, thus making the arm breech-loading. As most of our readers will have handled one of these Colt's pistols, it will be unnecessary to describe them; besides, the complicated nature of the mechanism would render any detailed account impossible without diagrams. This arm is fired by percussion caps; and the round bullet, rather larger than the bore of the barrel, takes the grooves while being pressed through it. Colt's invention having become popular, a great number of revolving small-arms have been invented, but only Deane and Adams have really simplified and improved it as an arm of war. Still, the whole thing is extremely complicated, and applicable, for war purposes, to pistols only. But, with a few improvements, this revolver will become a necessity for all cavalry, and for sailors when boarding, while for artillery it will be far more useful than any carbine. As it is, its effects at close quarters are terrible; and not only have the American cavalry been provided with them, but they have also been introduced into the British, American, French, Russian, and other navies.

The Swedish gun, as well as the revolver, are fired from without by common percussion caps. The third class of breech-loaders, the much talked-of Prussian needle-gun, does entirely away with these, too; the charge is fired from within.

The needle-gun was invented by a civilian, Mr. Dreyse, of Sömmerda, in Prussia. After having first invented the method of firing a gun by means of a needle suddenly penetrating an explosive substance fixed in the cartridge, he completed his invention, as early as 1835, by constructing a breech-loader, supplied with this needle-firing apparatus. The Prussian Government at once bought up the secret, and succeeded in keeping it to themselves up to 1848, when it became public; in the meantime they resolved upon giving this arm, in case of war, to all their infantry, and commenced manufacturing needle-guns. At present, the whole infantry of the line, and the greater portion of the Landwehr are armed with it, while all the light cavalry are at this moment receiving breech-loading needle-carbines.

Of the breech-loading mechanism we will only say that it seems to be the simplest, handiest, and most durable of all those that have, so far, been proposed. It has now been tried for years, and the only fault that can be found with it is this, that it does not last quite so long, and will not bear quite so many rounds as the fixed breech of a muzzle-loader. But this is a fault which appears unavoidable in all breech-loaders, and the necessity of renewing, a little sooner than with the old arms, a few pieces of the breech, cannot in any way detract from the great merits of the arm.

The cartridge contains bullet, powder, and the explosive composition, and is placed, unopened, into the chamber, which is slightly wider than the rifled barrel. A simple motion of the hand closes the breech, and at the same time cocks the gun. There is, however, no cock outside. Behind the charge, in a hollow iron cylinder, lies a strong, pointed steel needle, acted upon by a spiral spring. The cocking of the gun consists in merely drawing back, compressing, and holding fast this spring; when the trigger is drawn, it sets the spring loose, which at once sends the needle quickly forward into the cartridge, which it pierces, instantaneously explodes the explosive composition, and thus fires the charge. Thus, loading and firing with this gun consists of five motions only: opening the breech, placing the cartridge in it, closing the breech, presenting, and firing. No wonder that, with such a gun, five well-aimed rounds can be fired in a minute.

The projectiles first used for the needle-gun had a very unfavourable shape, and, consequently, gave a very high trajectory. This defect has been very successfully remedied a short time ago. The shot is now much longer, and has the shape of an acorn taken from its cup. It is of considerably smaller diameter than that of the bore; its rear-end is embedded in a kind of cup, of a soft material, so as to give it the requisite thickness. This cup sticks on to the bullet while in the barrel, takes the rifling, and thus gives the shot the lateral rotation, while at the same time it considerably diminishes friction in the barrel, and yet does away with all windage. The practice of the gun has been so much improved thereby, that the same sight, which formerly served for 600 paces (500 yards), now serves for 900 (750 yards); certainly an immense lowering of the trajectory.

Nothing is further from the truth than that the needle-gun is of a very complicated construction. The pieces composing the breech-loading apparatus and the needle-lock are not only far less numerous, but also far stronger than those composing a common percussion-lock, which yet nobody thinks too intricate for war and rough usage. Moreover, while the taking to pieces of a common percussion-lock is an affair requiring considerable time and sundry instruments, a needle-lock can be taken to pieces and refitted in an incredibly short time, and with no other instruments than the soldier's ten fingers. The only piece liable to break is the needle itself. But every soldier carries a reserve-needle, which he can fit to the lock at once, without having to take it to pieces, and even during an action. We are also informed that Mr. Dreyse has rendered the breaking of the needle a very unlikely thing, by an improvement in the lock, which makes the needle go back to its sheltered position as soon as it has done its work of exploding the charge.

The trajectory of the present Prussian needle-gun will be about the same as that of the Enfield rifle; its calibre is a little larger than that of the Enfield. With a reduction of calibre to that of the Austrian, or better still, the Swiss sharpshooter's rifle, there is no doubt that it would equal any of these arms in range, precision, and flatness of trajectory, while its other enormous advantages would remain to it. The breech-loading apparatus could even be made much stronger than at present, and the centre of gravity of the gun would be brought still nearer to the shoulder of the aiming soldier.

The introduction into an army of an arm capable of such rapid firing will necessarily produce many speculations as to what changes this will produce in tactics; especially among people so fond of speculating as the North Germans. There has been no end of controversies on the pretended revolution in tactics which the needle-gun was to produce. The majority of the military public, in Prussia, at last came to the result that no charge could be made against a battalion firing needle-gun volleys in rapid succeession, and that consequently it was all up with the bayonet. If this foolish notion had prevailed, the needle-gun would have brought upon the Prussians many a severe defeat. Fortunately, the Italian war proved to all who could see, that the fire from modern rifles is not necessarily so very dangerous to a battalion charging with spirit, and Prince Frederick Charles of Prussia has taken occasion therefrom to remind his comrades that passive defence, if ever so well armed, is always sure of defeat. The tide of military opinion has turned. People again begin to see that men, and not muskets, must win battles; and if any real change in tactics will be made by the new gun, it will be a return to a greater use of deployed lines (where the ground admits of it), and even to that charge in line which, after having won most of the battles of Frederick the Great, had become almost unknown to the Prussian infantry.

17. The History of the Rifle[1]—VII

Having now passed in review the different systems upon which the various rifles, now in use in European armies, are constructed, we cannot take leave of our subject without saying a few words with respect to a rifle which, although now introduced into any service, enjoys a well-deserved popularity for its astonishing precision at long ranges. We mean, of course, the Whitworth rifle.

Mr. Whitworth, if we are not mistaken, claims as original two principles in the construction of his fire-arms—the hexagonal bore and the mechanical fit of the projectile in the bore. The bore, instead of having a circular, has a hexagonal section throughout, and a very strong pitch or turn, as is shown on the surface of one of the hexagonal bullets. The bullet itself is of a hard metal, fits the bore as nice as possible, and is not expected to alter its shape in consequence of the explosion, as its six corners make it follow the twist of the grooves with unerring certainty. To prevent windage, and to lubricate the bore, a cake or bottom of greasy matter is inserted between the powder and the charge; this grease melts from the heat of the explosion, while travelling, behind the bullet, towards the muzzle.

Now, in spite of the undeniably excellent results which Mr. Whitworth has obtained with his rifle, we believe that this principle is inferior to either that of expansion, or of compression, or of breech-loading with a bullet larger in diameter than the bore. That is to say, we believe that either a rifle for expansion-shot, or one for compression-shot, or one constructed on the system of the Prussian needle-gun, would beat a Whitworth rifle if the workmanship was equally good, the calibre equally small, and all other circumstances alike. Mr. Whitworth's mechanical fit may be ever so nice, he cannot make it as close as the change in the shape of the bullet during and after the explosion makes it. There is in his rifles with hard bullets always that which a rifle is meant radically to avoid, namely, windage and consequent escape of gas; even the melting grease cannot entirely do away with that, especially in a rifle which, from long use, has become a trifle larger in the bore. There is a very distinct limit to all mechanical fit in such a case, and that is, the fit must be loose enough to let

[[1] *Volunteer Journal*, Vol. I, No. 19, pp. 247-8, January 12, 1861. Article by the author of 'A German Account of the Newton Review'.]

the bullet go down easily and quickly, even after a couple of dozen rounds. The consequence is that these hexagonal bullets do fit but loosely, and although we do not know exactly what the amount of windage is, still the fact that they will go down quite easily without any grease and with a piece of paper wrapped round them, makes it probable that it is not much less (if less at all) than that of the Enfield bullet, which is one-hundredth part of an inch. Mr. Whitworth, in contriving this rifle, seems to have had chiefly two leading ideas: firstly, to do away with all possibility of getting the grooves loaded; and, secondly, to do away with all the accidents which may prevent a cylindrical bullet from taking the rifling—because they prevent either expansion or compression taking place—by adapting the shape of the bore and that of the shot to each other beforehand. The obstruction of the grooves by particles of lead torn off from the bullet may occur in all rifles with soft leaden bullets; the accidents preventing a bullet from taking the grooves in the correct way may occur in either compression or expansion rifles, but not in breech-loaders on the Prussian principle. But neither of these inconveniences is so great that they cannot be overcome, and that, in order to avoid them, the first principle in rifle making should be sacrificed, viz., that the bullet takes the rifling without leaving any windage.

In saying so, we are backed by an excellent authority, namely by Mr. Whitworth himself. We are informed that Mr. Whitworth has dropped his principle of mechanical fit as far as his rifle is concerned, and certain it is that at present most people fire from his rifle not a hard, solid, hexagonal bullet, but a soft, leaden, cylindrical bullet. This bullet is hollowed out at its base similar to the Enfield bullet, but it has no plug; it is very long (the one 480 grains, three times as long as its diameter, the other, 530 grains, three and a half times its diameter), and takes the rifling by the effect of the explosion. Here, then, we have Mr. Whitworth's principle of mechanical fit entirely abandoned for that of expansion, and the Whitworth rifle turned into a subordinate species of the genus Minié quite as much as the Enfield ever was. Remains the hexagonal bore; and how will that answer for an expansion rifle?

The hexagonal bore has, of course, six grooves, and we have seen that an even number of grooves has been found to answer, for expansion bullets, not so well as an uneven one, as it is not desirable that two grooves should be diametrically opposite to each other. Then the grooves in most expansion-rifles are very shallow—in the Enfield, for instance, scarcely visible. In the hexagon the difference between the diameter of the inner circle (representing the bore at large) and that of the outer circle (drawn through the six corners) is about $\frac{2}{13}$ths, or rather less than one-sixth part of the former; or, in other words, the lead has to expand nearly one-sixth of its diameter before it can properly close to the corners of the hexagonal bore. From this it would appear that the hexagonal bore, although exceedingly ingenious for the system of mechanical fit, is about the most unlikely to answer for the system of expansion.

Still it answers, as the results of almost every rifle contest prove. How is this possible, if Mr. Whitworth has abandoned the essential point of his principle, and now applies a principle for which his rifle is not adapted?

First of all, there is the excellence of the workmanship. It is well known that for accuracy in the most minute and even micrometrical details, Mr. Whitworth stands unrivalled. As his engineering tools, so are his rifles; perfect models in the construction of their detail. Look at the sight on the muzzle of his rifles, and at that of any other class! There is no comparison: and in rifles firing at 1,000 yards range, this is an immense advantage.

Secondly, and chiefly: the calibre of the Whitworth rifle is 0·451 of an inch minimum bore (what we have called the inner circle). The Enfield is 0·577; the Swiss sharpshooter's rifle, which we have more than once mentioned as giving the lowest trajectory known, is 0·413. Now, look at the difference in the shape of the bullet. The Whitworth expansion bullet of 530 grains is about three-eights of an inch longer than the Enfield bullet of the same weight; while the former is about three and a half times its own diameter in length, the latter is scarcely twice its own diameter. It is evident that a bullet of the same weight and with the same charge will cut better through the air, that is, give a lower trajectory if it is thin and long, than if it is short and thick. Then, the charge of the Enfield is 68 grains of powder; for the Whitworth, charges of 60, 70 and 80 grains are used, but we have been told by good shots who are in the habit of using this rifle that 80 grains are required to make the bullet expand well and give good results at long ranges. Thus we have a charge for the Whitworth fully one-sixth stronger than for the Enfield, and that charge would act better (even with equal weight), as it explodes in a more confined space and acts upon a far smaller surface of bullet.

Here, then, we have another specimen of the immense advantage of a small bore, which gives a long, thin, bolt-shaped shot. Whoever of my readers has attentively followed our inquiries into the advantages of the various rifles, will have long since come to the conclusion that the shape of the bullet is of far more importance than the system on which either shot or rifle are designed; and that in order to have a portable soldier's bullet of the best shape, we must have a small bore. This is the lesson the Whitworth rifle again teaches us.

We may also learn from it that, with a small bore, the long, heavy point of the bullet offers resistance enough to allow the hollow tail end to expand with certainty, and without the assistance of a plug. The Whitworth bullet has but a small cavity at its base, and no plug; it has to expand at least three times as much as any other expansion bullet; and still, with 80 grains of powder (which the rifle stands without too much kicking), it does take the rifling quite sufficiently.

That Mr. Whitworth's rifle will ever become a weapon of war, we very much doubt; indeed, we think the hexagonal bore will soon go out altogether. If volunteers who had become practically convinced of the superior shooting of the Whitworth rifle as compared with the present Enfield, have proposed that they should be armed with the former, they have certainly far overshot the mark. We think it utterly unfair to compare the two species of arms. The Whitworth is an arm of luxury, which costs at least twice as much as the Enfield to produce. In its present state it is too delicate a weapon to be placed into every soldier's hands; but take, for instance, the delicate sight from the muzzle, replacing it by one fit for rough usage, and its accuracy at long ranges will be considerably diminished. To arm both army and volunteers with the Whitworth, one of two things must be done; either the calibre of the regulation small-arms must remain the same as now, and then a Whitworth, with the bore of the present Enfield, would give far worse results than the present Whitworth, or the bore must be reduced, say to that of the present Whitworth, and then it is probable that an Enfield with that reduced bore, on the making of which as much had been spent as on a Whitworth, would give as good or better results.

18. The History of the Rifle[1]—VIII

We conclude with a short recapitulation of the different systems of rifles now in use, and of the principles which we may consider as established with regard to this arm.

The different systems of rifles are as follows:

1. The system of forcible loading, the tight fitting bullet and plaster being hammered down by strong blows of the ramrod. This is the oldest plan of making a bullet take the rifling. It has now been almost universally abandoned for arms of war; the principal and very remarkable exception being the new Swiss sharpshooter's rifle, which has a very small calibre and a long, bolt-shaped shot, and which gives, of all rifles now in use, the lowest trajectory. It is not intended for an arm for the mass of the infantry, but for select bodies only, and requires careful loading in order to give the highly favourable results which distinguish it above all other rifles now known.

2. The system of flattening the loose fitting bullet against some obstacle at the bottom of the breech (either the rim of a narrow chamber—Delvigne—or a peg placed in the middle of the chamber—Thouvenin) and thus drive it into the grooves. This plan, for a time very generally favoured, is now becoming more or less superseded by the following systems. Let us observe, at the same time, that it requires a rather large calibre, as otherwise the chamber becomes too narrow.

3. The system of expansion, the loose fitting, elongated shot being hollowed out from the base, and the gas created by the explosion entering into the cavity and blowing it up, so to say, to a sufficient degree to make the bullet fit the bore and take the rifling. This system now is in general favour, and is still capable of great improvement, as has latterly been shown by the explosion entering into the cavity and blowing it up, so to say,to a sufficient degree to make the bullet fit the bore and take the rifling. This system now is in general favour, and is still capable of great improvement, as has latterly been shown by the excellent result

[1 *Volunteer Journal*, Vol. I, No. 20, pp. 259–60, January 19, 1861. Article by the author of 'A German Account of the Newton Review'.]

which Mr. Whitworth obtained with his rifle since he adopted the principle of expansion.

4. The system of compression, in which the same result is obtained by providing the bullets with deep, circular indentations, which allow the explosive force, while opposed by the weight of the heavy fore part of the projectile, to compress it lengthways, and thereby give it the required increase of diameter. This plan, although evidently less safe than the expansion principle, has given excellent results with small calibres, as has been proved in Austria and Switzerland. Still, the compression-bullet, fired from the Swiss sharpshooter's rifle above alluded to, does not give quite as good results as the tight-fitting plaster bullet from the same arm.

5. The breech-loading system, which has advantages of its own over all other systems of rifles in the mode of loading and firing, offers, at the same time, the greatest certainty of the bullet taking the rifling, as the chamber and bullet may be made slightly larger than the rest of the bore, and thus the bullet cannot get to the muzzle without being pressed into the grooves. This system, indeed, appears to be destined gradually to supersede all other systems.

We do not count Mr. Whitworth's system of mechanical fit, as it has been abandoned as far, at least, as small arms are concerned; and with these alone we have now to do.

If the various systems are classed according to their intrinsic merits, we should say that the breech-loading needle-gun stands highest; next, the expansion system; then the compression system. The two first systems may be considered to be superseded; for even if forcible loading, in Switzerland, so far gives better results, with the same calibre, than compression, we should not at all be inclined to give to the system the credit of these results without a very searching examination; and, besides, the Swiss sharpshooters' plaster bullet is acknowledged to be unfit for the mass of the infantry.

At the same time, we have seen that since the introduction of elongated bullets, the system on which either rifle or shot is constructed is of but secondary importance in obtaining great range, low trajectory, and accuracy of flight. As long as bullets were round, the system of rifling was of greater importance, for then all bullets were met by the resistance of the air under nearly equal circumstances, and the influence of a stronger pitch of rifling, of deeper or more numerous grooves, etc., were comparatively far more important than now. But with elongated shot, a new element appears on the ground. The bullet may be made longer or shorter, within pretty wide limits, and now the question is which shape of bullet is most advantageous? On theoretical grounds it is clear that the same mass of lead, started with the same initial velocity, will better retain that velocity if its shape is long and thin, than if it is short and thick; supposing always that the lateral rotation which a rifle would give it, is there to prevent its going head over heels. The resistance of the air is the retarding force; it gradually diminishes the original velocity imparted to the bullet by the powder, and thus gives the ever-increasing force of gravity, so to speak, a greater hold upon the projectile. The initial velocity depends upon the charge, and in some degree upon the construction of the arm; this we may, therefore, consider to be fixed; the force of gravity is also fixed, and a given quantity; remains, as variable, the shape of the bullet to enable it to dart through the air with the least amount of resistance; and to evade atmospheric resistance, as we have said, a long and thin shot is far better fitted than a short and thick one of the same weight.

Now, the maximum weight of the bullet for military purposes is also a given quantity. A man must be able to carry, at least, sixty rounds over and above his arms and accoutrements. To produce the best-shaped bullet, therefore, out of this given weight of lead (say 530 grains), the length must be increased and the thickness diminished;—in other words, the bore of the rifle must be made less. Up to a certain point this will hold good without exception. Look at the 530 grains in the Enfield and at the same weight in the Whitworth bullet; a single glance explains why the latter has a so much lower trajectory (that is, retains its initial velocity so much better, and will, therefore, hit a target at a 1,000 yards with ease, while the Enfield cannot be trusted at that distance. And yet, the two are both expansion bullets, and the general construction of the Whitworth is certainly not the best adapted for expansion. Or look at the Swiss sharpshooter's rifle, with a bore still smaller than the Whitworth, and giving still better results and a still lower trajectory, be its bullet rammed home with a plaster, or let down loosely and compressed by the explosion. Or take the Prussian needle-gun; by reducing the diameter and increasing the length of the bullet, and guiding it in the wide bore by a button or wad, the same sight which formerly marked the 600 yards' range, now carries the bullet to 900 yards, we shall, therefore, be pretty safe in considering it as an established fact that, in a general way, the efficiency of rifles, no matter on what system they are constructed, will be in the inverse ratio of the diameters of their bores. The smaller the bore, the better the rifle, and vice versa.

With these observations we take leave of a subject which may have appeared rather dry to many of our readers. Still, its importance is very great. No intelligent soldier ought to be ignorant of the principles on which his arms are constructed and are expected to act. What we have attempted to expose here, the non-commissioned officers of most Continental armies are expected to know; and surely, the majority of the volunteers, 'the intelligence of the country', ought to be as well up in the knowledge of their firearms as they!

19. Rifles and Rifle-Shooting: The Lancaster and Enfield Rifles[1]

The recent contest between Lieut. Wallinger and the sergeants of the Royal Engineers, reported in our numbers for April 6th and 13th, has recalled public attention to the merits of the Lancaster rifle, especially as a service weapon. In the match at Chatham the sergeants fired with the ordinary military ·577 oval-bore Lancaster carbine of the Royal Engineers, the cost of which is about £4. To match such a weapon with the highly-finished Whitworth, costing about £25, is evidently unfair. A more equal comparison might be instituted between the Lancaster and the ordinary Enfield, because the difference in the cost of these two weapons is not very material, and the price of the Lancaster would probably be reduced to an equality with the Enfield if it were manufactured in as great numbers at the Government factories. The question then remains, is it a better rifle? A writer in the *London Review*, reasoning from general principles, and judging also from actual experience, answers in the affirmative; and

[[1] *Volunteer Journal*, Vol. II, No. 35, pp. 85–6, May 4, 1861.]

we invite attention to the following passages from his article on the subject:

'The law which governs accurate rifle-shooting or practice is very simple. It is only necessary to establish an equation between the length and diameter of the ball, and to give to that ball an adequate rotatory motion around its polar axis, when unfailing accuracy must be the result, irrespective of the precise method by which the rotatory or rifle motion is given. That is to say, the interior of the rifle barrel may be cut into any number or any shape of groove, or no grooves at all, so long as the equation is preserved, and the bullet acquires a proper rotatory motion, and then the accuracy in every case will be equal. The consideration, however, that must determine the proper arm for a soldier involves as first conditions that the weapon should not exceed a certain weight and dimension, and that it should be easily loaded and easily cleaned. It therefore follows, that to be easily loaded, the bearing surface, in the act of loading, should be as small as possible; and that, in the shape given to the rifling, as far as practicable, all angles should be avoided. We know no other form that so perfectly carries out this proposition as the spiral oval, inasmuch as the bearing surfaces in the act of loading are but two, and no form offers so great facilities for cleaning with the unavoidably scanty means at the disposition of the soldier during active service. This opinion seems to be borne out by the results of the Indian campaign, and by the trials at Malta, Gibraltar, and other foreign stations. In India the Enfield rifle is said to have completely "shut up" at many critical periods of the campaign. The papers and private letters and official reports teemed with complaints; yet with the same ammunition, under the same circumstances, the oval-bore rifles with which the Royal Engineers were armed never failed to perform their duty to the satisfaction of both officers and men.

'When the Enfield rifle is made with a diminished bore and an elongated bullet is used, comparably with the Whitworth the effect is just as good; yet the Enfield service-rifle, as it now exists, must be regarded as an attempt to satisfy impossible conditions. The officers charged with the construction of this arm were not permitted to reduce the calibre of the weapon below a given limit. Hence the adoption of the standard bore of

·577. As a consequence of this too great diameter of bore, an inherent difficulty presented itself, namely, that of securing a perfectly and unfailing hermetical fit between the interior of the bore and the ball when driven from the barrel by the explosion of the powder. Let us examine the actual result of the imperfect conditions exhibited in the Enfield rifle. The weight of the ball is fixed at 530 grains, the charge of powder at 70 grains, the calibre, as before stated, at ·577. Now, the effect of 70 grains of powder acting on the large cross-section of the ball, will not and does not give pressure sufficient to produce in every case sufficient expansion of the ball into the grooves. Careful experiment shows that not 10 per cent. of the bullets are equally and fully expanded on every side. Sometimes one groove is distinctly marked, sometimes two, and in only one-tenth of the total rounds are they fully expanded, hence the inaccuracy of the shooting of the ·577 bore service-rifle.

'Now, the perfect conditions of accurate practice from rifles grooved in *any form* may be described as follows: That the bore should be ·5 inch, the length of ball 1·12 inch, rotation or twist 1 in 18 inches, charge of powder 90 to 100 grains (No. 6), weight of ball identical, namely 530 grains. The force exerted under this condition upon the cross-section of the ball may be considered as plus, therefore there is an unfailing and unerring fit between

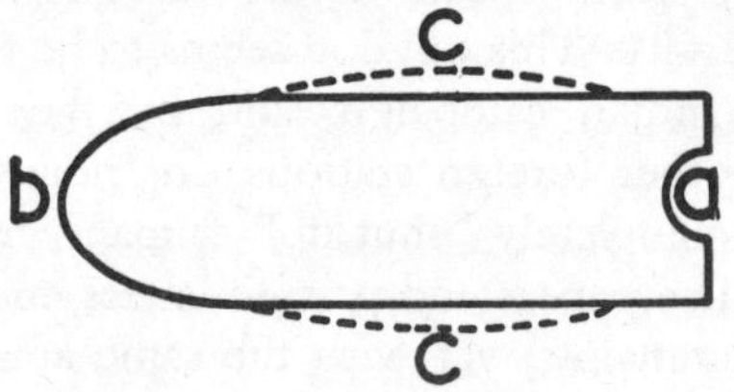

BULLET FOR SMALL-BORE RIFLE

the ball and the bore, and it arises in this way: the diminished diameter of the bore gives increased length of ball, and no wooden plug is necessary as in the service bullets to drive out the metal. The bullet is therefore an homogeneous solid of about three diameters long. In the explosion the expansive force of the powder is first exerted on the rear or posterior section of the ball (*a*), and the transmission of the motive force, although almost instantaneous; is nevertheless met by the *vis inertiae* of the mass of metal constituting the ball, exerted in the whole length (from *a* to (*b*), and backed by the counter resistance of the air in the barrel. It is at a glance evident that this resistance must be evinced in the middle portion of the bullet or part of greatest resistance (*c*), and consequently, by a perfectly natural expansion there, the bullet would be slightly shortened, say about one-tenth of an inch, while the central portion would be increased in diameter sufficiently to fit hermetically the form of the interior of the barrel, whatever its shape might be.

'When these more perfect conditions are fulfilled, not once in 500 times is there any defective expansion, the ball invariably taking the form of the rifle, and thus there results the most excellent rifle practice.

'These remarks apply to all rifles of every description.

'What is it these favourable conditions do for a rifle, and why do they give more accurate shooting? Having shown how the interior of the bore is perfectly fitted by the bullet, we will endeavour to trace its results. One of the main achievements in the construction of a rifle is to get a "low trajectory," that is, that the curve the bullet describes in its flight should be as near an approach to a straight line as possible, and, as a necessary sequence, a high velocity is absolutely necessary, so that gravitation may have the minimum of effort in depressing the ball in its passage. Now, the effect of diminution of the calibre achieves the first result, and by the employment of a larger charge of powder on the small cross-section of the ball, the highest velocity and the most accurate results are obtained.

'With respect to the methods of rifling, it will be inferred from what we have already said that so long as the ball gets a proper "spin" on leaving the barrel, it matters not in itself how that rotation is given, whether by an hexagonal bore, as in the Whitworth, an oval, as in the Lancaster, or by three grooves, as in the Enfield. Neither is a number of grooves necessary, for if *one* has a sufficient grip on the bullet to turn it, the requisite condition is fulfilled. Still, there are inherent defects in the manners of grooving which may easily be shown. If the rifling be angular, there is a loss of power in effecting the expansion necessary to fill up the angles, besides the probable escape there of the propelling gas. Moreover each angle is a line of weakness to the barrel; so with any number of grooves, and proportionably to their depths are the same defects manifested. The spiral oval, therefore, of the Lancaster gun is theoretically the best, as presenting the form to which the bullet will most readily adapt itself with the slightest expansion.

'That the Lancaster rifle must have great merits appears from the fact, that, before the adoption of the Enfield pattern, the Lancaster rifle, then in competition with it, was recommended in preference by four separate and distinct committees. It was submitted for approval to the Commander-in-Chief, and by him sent for final decision to Hythe. The first reports from the officers of the School of Musketry there was most favourable; the second report decided in favour of the Enfield. The reason then assigned for this decision was that the balls "stripped". Subsequently, however, the following facts are said to have transpired. The first 10,000 rounds of Pritchett ammunition, with which the first trials there were conducted, were of the proper standard diameter. With these cartridges admirable shooting resulted. In the second experiment the same ammunition was not employed, the former having been made in 1853, the latter in 1854; the experimenting officers at Hythe being entirely unconscious of any difference in the ammunition, not having been informed that in the bullets made in the latter year, there was a difference of ·007 less in the diameter, as compared with the bullet of 1853.

'This fact was not detected until a year and a half after the final decision in favour of the Enfield, when Colonel (then Captain) Fitzroy Somerset, tested the pattern of the Royal Engineer oval-bore carbine. It is easy to see that the diameter of the diminished Pritchett bullet being less than the proper standard, it would, in many instances, especially when there was any excess of hardness in the lead, pass out of the barrel without acquiring a rotatory motion, that is, it would not sufficiently expand to fill the interior of the barrel, whether of a Lancaster or any other rifle.

'That the Whitworth is too expensive for army use, and requires more delicate treatment than it is likely to get in actual service, we think few will question, and the tests should therefore be applied to the Lancaster and Enfield or other patterns respectively, which are fitted for the rough usage of warfare. The way of doing this, however, will not be by rifle-matches, but by firing from a fixed rest, with equal quantities of powder, and with bullets of the same weight and cast, thus making all such conditions equal, and leaving the test open only to the respective merits of the weapons themselves.'

The preceding remarks refer to two different questions: 1. Which is the best proportion between the diameter and the length of an elongated rifle-shot to be fired from *any* rifle? And, 2. What are the merits of the Lancaster or oval-bore rifle?

As to question No. 1, we are far from agreeing with the author, that the proportions of his best bullet are preferable to all others. The rifles which, so far, have given the best results—the Swiss and the Whitworth—have both a smaller calibre than 0·5, and a greater proportional length of shot. We cannot, however, here enter into a discussion on a point of such a general nature.

As to question No. 2, we cannot see what positive evidence the author gives of *any* superiority of the Lancaster rifle over the Enfield. That the carbines of the Engineers 'shut up' less often than the Enfield rifles of the infantry, is easily explained by the fact that the infantry are a hundred times more numerous, in any army, than the Engineers; and that the latter do not use their carbines once when the line use their rifles a hundred times; because Engineers are there for other purposes altogether than to act as infantry.

That a long and heavy expansion-shot, hollowed out sufficiently at the rear end, with a full charge, can be made to take almost any shape of rifling, is proved in the instance of the Whitworth; here the amount of expansion required is extremely great, and still the bullet takes the hexagonal shape at its rear end. No doubt, therefore, such a bullet can be made to expand sufficiently to fill up an oval bore, if the difference of the two diameters be not too great. But why on that account the Engineer carbine should be better than the Enfield is more than we can perceive. The ideal bullet of our author has nothing whatever to do with this carbine—it would not fit it; and if even with a reduction of calibre, our author considers an increased charge of 90 to 100 grains of powder necessary to make his bullet fully take the oval bore, we think that looks much like a silent admission that the present charge of 70 grains does *not* always ensure a full expansion of the bullet in the oval bore of the Engineers' carbine. Our author does not say what is to become of the increased recoil from the increased charge; still we know that 80 to 90 grains give, in the Whitworth, a not very pleasant amount of recoil, which, in rapid firing, very soon affects the steadiness of aim.

The uncommonly good results given by the Engineers' carbine in the Chatham match, as well as some exceedingly good shooting with Lancaster rifles by private gentlemen, mentioned at times in the press, make it desirable that the capabilities of the oval-bore expansion rifle, and its fitness for a service weapon, should again be tried. We, for our part, believe that it will be found to have its faults too, and that the principle of the rifling is a very secondary matter indeed in military muskets. Instead of quibbling with the Enfield about such minor matters, why not come to the point at once, and say that its greatest and most important defect is its *large calibre*? Change that, and you will find all other improvements but matters of detail.

PART III

THE FRENCH ARMY

20. The French Light Infantry—I[1]

If ever our volunteers should have to exchange bullets with an enemy, that enemy will be—everybody knows it—French infantry; and the finest type, the *beau idéal* of a French foot-soldier, is the light infantry soldier, especially the *chasseur*.

The French chasseur is not only the model for his own army; the French give the law, to a certain degree, to all European armies in matters regarding light infantry service; thus the chasseur becomes, in a certain sense, a model for all European light infantry.

In both these qualities, as a possible opponent, and as, hitherto, the most perfect specimen of a light infantry soldier, the French chasseur is a subject of high interest to the British volunteer. The sooner our volunteer gets acquainted with him the better.

Up to 1838, there was not a rifle in use in the French army. The old rifle, with its close-fitting bullet, which had to be hammered down, and made loading a difficult and slow operation, was no arm for the French. When Napoleon once examined the firelocks of a German battalion of rifles, he exclaimed: 'Surely this is the most unfortunate arm to give into the hands of a soldier.' The old rifle was, certainly, unfit for the great mass of the infantry. In Germany and Switzerland, a few chosen battalions were always armed with it, but they were exclusively used as sharpshooters, to pick off officers, to fire on sappers constructing a bridge, etc.; and great care was taken to form these corps from the sons of gamekeepers, or other young men who had been trained to the use of the rifle long before they entered the army. The chamois-hunters of the Alps, the keepers of the great deer forests of Northern Germany, formed excellent material for

[1 *Volunteer Journal*, Vol. I, No. 3, p. 42, September 21, 1860.]

these battalions, and they too, were the model for the rifles of the English line.

What the French formerly used to call Light Infantry, were men equipped and drilled exactly the same as the regiments of the line; consequently, in 1854, a decree of Louis Napoleon deprived these 25 regiments of the name of Light Infantry, and embodied them in the line where they now number from the 76th to the 100th regiment.

There was, indeed, in every battalion of infantry a company of *voltigeurs*, formed of the best and most intelligent soldiers of small stature; the *élite* of the taller men being formed into the company of grenadiers. They are the first to extend when skirmishers are required, but in every other respect they are armed and drilled like the remainder of the battalion.

After the conquest of Algiers, in 1830, the French found themselves face to face with an enemy armed with the long musket, common to most Eastern nations. The smooth-bore muskets of the French were inferior to them in range. The French columns on the march were surrounded on every side by mounted Bedouins in the plains, by Kabyle skirmishers in the mountains; the bullets of these enemies told on the columns, while they themselves were out of effective range of the French fire. Skirmishers, in the plains, could not move far from their columns, for fear of being surprised and cut up by the rapid Arab horsemen.

When the English army got into Afghanistan, it made acquaintance with these same long muskets. The Afghan shots—though from matchlocks only—did fearful execution in the English ranks, both in the camp at Kabul and during the retreat through the hills, at distances utterly unattainable to poor old Brown Bess. The lesson was a severe one; the war was to be renewed; protracted conflicts with the tribes on the north-eastern frontier of British India might be expected; yet nothing was done to arm the English soldiers sent to that frontier with a weapon able to cope at long range with the Eastern matchlock.

Not so the French. No sooner was the defect found out than steps were taken to remedy it. The Duke of Orleans, the son of Louis Philippe, on his matrimonial tour through Germany in 1836, took occasion to study the organization of the two battalions of rifles of the Prussian guard. He saw at once that here was a starting point, issuing from which, he might succeed in forming the very class of troops required for Algeria. He occupied himself at once with the subject. The old French prejudice against the rifle placed many obstacles in his way. Fortunately, the inventions of Delvigne and Pontchara, in his own country, came to his help; they had produced a rifle which could be loaded almost as quickly and easily as the smooth-bore musket, while it exceeded the latter by far, both in range and precision. In 1838, the Duke obtained permission to form a company according to his own ideas; in the same year, this company was increased to a full battalion; in 1840, it was sent to Algeria to prove what it could do in actual war; and it stood the test so well, that in the same year nine more battalions of Chasseurs were formed. Finally, in 1853, ten of the battalions were organized, so that the whole Chasseur force of the French army now consists of twenty battalions.

The peculiar military qualities of the Bedouins and Kabyles, who undoubtedly were models of light horsemen and of infantry skirmishers, very soon induced the French to try the enlistment of natives, and to conquer Algeria by setting Arab to fight Arab. This idea gave origin, among others, to the corps of the *Zouaves*. They were formed, principally of natives, as early as 1830, and remained a chiefly Arab corps up to 1839, when they deserted in masses into the camp of Abd-el-Kader, who had just raised the standard of holy war. There remained then merely the cadres and the twelve French soldiers of each company, besides the two exclusively French companies attached to each battalion. The vacancies had to be filled up by Frenchmen, and since that date the Zouaves have remained an exclusively French corps, destined to take permanent garrison in Africa. But the original stock of old French Zouaves had adopted so much of the native character, that the whole corps has ever since remained, in its entire spirit and habits, a specially Algerian corps, endowed with a nationality of its own, and quite distinct from the remainder of the French army. They are recruited mostly from substitutes, and thus they are most of them professional soldiers for life. They, too, essentially belong to the Light Infantry of the army and have, therefore, been long since provided with rifles. There are now three regiments or nine battalions of them in Africa, and one regiment (two battalions) of Zouaves of the Guard.

Since 1841, new attempts were made to enlist native Algerians for the local army. Three battalions were formed, but they remained weak and incomplete till 1852, when more encouragement was give to native enlistment; and this succeeded so far that, in 1855, three regiments, or nine battalions, could be formed. These are the *Turcos* or *Tirailleurs indigènes*, of whom we have heard so much during the Crimean and Italian wars.

Thus, not counting the foreign legion (now disbanded, but to all appearances re-forming) and the three penal battalions, the French army contains 38 battalions, especially formed and trained for light service. Of these, the Chasseurs, the Zouaves, and the Turcos, each have their distinguishing characteristics. Troops like the last two classes have too strongly marked a local character ever to exercise a great influence upon the mass of the French army; still, their furious onslaught—during which they still, as has been proved in Italy, remain perfectly in hand, and even anticipate, by their own military tact, the orders of their chief,—will always remain a brilliant example to the remainder of the troops. It is also a fact that the French, in their practice of the detail of skirmishing, and their mode of taking advantage of ground, have adopted a great deal from the Arabs. But that class of light infantry which has remained essentially French, and has thereby become, as we said before, a model to the army, are the Chasseurs, of whom more in our next number.

21. THE FRENCH LIGHT INFANTRY—II: THE CHASSEURS[1]

The very first page of the French Drill Regulations of 1831, proves what little men the French army is composed of.

'Slow time, each step 65 centimetres (25 inches, and 76 paces in a minute.

'Quick time, same length of step, and 100 paces in a minute.

'Charging time (*pas de charge*), same length of step, and 130 paces in a minute.'

The step of 25 inches is undoubtedly the shortest, and the celerity of 100 paces in a minute the most sluggish adopted in an army for field-movements. While a French battalion moves

[[1] *Volunteer Journal*, Vol. I, No. 5, pp. 73–4, October 5, 1860. Article by the author of 'A German Account of the Newton Review'.]

over 208 feet of ground in a minute, an English, Prussian, or Austrian battalion would move over 270 feet, or thirty per cent more. Our long step of 30 inches would be too much for the short legs of Frenchmen. The same at a charge: the French advance, in a minute, 271 feet, or as much as the English at simple quick time, while the English, at their double of 36 inches, and 150 per minute, would get over 450 feet, or sixty per cent more. This fact alone shows that the standard size of the men cannot be reduced beyond a certain limit without affecting the efficiency and mobility of an army.

With such short legs, short steps, and slow marching time, no light infantry could be formed. When the Chasseurs were first organized, care was taken from the very beginning to select the best infantry material in the country; they were all well-built, broad-shouldered, active men, from 5 ft. 4 in. to 5 ft. 8 in. in height, and mostly chosen from the mountainous parts of the country. By the regulations for Chasseur drill and evolutions, (published in 1845), the length of the step for the quick march was retained, but the time increased to 110 in a minute; the double (*pas gymnastique*) was regulated at 33 inches, (83 centimetres) each step, and 165 in a minute; but for deployments, formation of square, or other urgent occasions its time is to be increased to 180 in a minute. Even at this latter pace, the Chasseur would cover but 45 feet more ground in a minute than the English soldier at his double. But it is less by extraordinary rapidity of motion that extraordinary results are attained, than by the length of time for which the Chasseurs can continue this accelerated motion; besides, in cases of great urgency, rallying, etc., they are ordered to run as fast as they can.

The double is the principal thing practised in the Chasseur battalions. The men are first taught to mark the time at 165 and 180 per minute, during which they shout One! Two! or Right! Left! which is supposed to regulate the action of the lungs, and to prevent inflammations. They are then made to march forward at the same rate, and the distance is gradually increased until they can go over a French league of 4,000 metres (two and a half miles), in twenty-seven minutes. If some of the recruits are found too weak in wind and limb for such exercise, they are sent back to the infantry of the line. The next step is the practice of leaping and running, in which latter pace the greatest possible rapidity has to be obtained for short distances, both the *pas gymnastique* and the running being practised first on the level drill-ground, or on the road, and afterwards across country, with jumping over rails and ditches. After such preparation only are the men entrusted with their arms, and now the whole course of double, running, and leaping is again gone through with rifle in hand, and in heavy marching order, the knapsack and pouch weighted to the same extent as in the field; and thus they are made to continue for a full hour at the *pas gymnastique*, during which time they have to cover at least five miles of ground. A foreign officer in plain clothes once attempted to keep pace with such a battalion of Chasseurs in heavy marching order; but, untrained as he was, he could scarcely keep up for one hour; the Chasseurs marched on, alternately at quick time and at the *pas gymnastique*, and went that day over 22 miles of country.

The whole of the field movements and evolutions have to be gone through at the double; advance in line, forming column and square, wheeling, deployments, and everything, so that the men keep in their places as steadily at this pace as at the ordinary quick time. The time for all evolutions is 165 in a minute, only in deployments and wheelings, it is accelerated to 180.

The following is the opinion of a Prussian field officer of the Chasseurs:

'On the Champ-de-Mars, I saw a few companies of Chasseurs manœuvring at the side of a regiment of the line. What a contrast, from their mobility, from the whole style of their movements, to that regiment! At the first glance you see that they are a picked body, chosen from the best men of the wood and mountain districts; they are all well-knit, compact, strong, and yet so wonderfully nimble. As they flit about with astonishing rapidity, you recognize their enterprising spirit, their daring pluck, their quick intellect, their indefatigable endurance, though, certainly, you also recognize their immense conceit and French vanity. And wherever you see them, in Strasbourg, in Paris, or in any other garrison, they always make the same impression, they look as if cast in the same mould. At their head I saw none but young officers; a few only of the captains appeared thirty-five; most of them less, and even the field-officers not older. Their rapid mobility shows neither constraint nor effort; constant exercise appears to have made it their second nature, with such ease and freedom do these battalions go through their movements. Their blood has a more tranquil flow, their breath is less disturbed than with others. Single orderlies in a street, would pass, in a short time, all persons walking before them; and at the same quick pace, whole battalions, at the merry sound of the bugle, defile through the streets. Whenever you see them, on the drill-ground, on the march out or home, never did they appear tired to me. Ambition, in this matter, may go hand in hand with habit.

'If quickness of motion and steadiness of aim appear to be irreconcilable, the Chasseurs seem to have overcome this apparent incompatibility. I have not myself seen them practising at the target; but, according to the judgment of experienced officers, their performances in this line are not to be thought little of. If their steadiness of aim is at all disturbed, it certainly must be so in a degree very little affecting their efficiency on the field of battle. In Africa, where many an engagement was preceded by similar marches at the double, they have always known how to hit their opponents; and this proves that the special system of training to which they are subjected, tends to properly develop the powers of the body, and does not destroy steadiness of aim. With troops not so trained, this would, of course, be very different.

'The great advantages of this system of training are evident. Many are the cases in war, in which it may be of decisive importance that your infantry should be capable of quicker locomotion than it is at present; for instance, in preceding the enemy in the occupation of an important position; in rapidly attaining a commanding point; in supporting a body attended by superior forces; or in surprising the enemy by making a detachment suddenly appear in a direction quite unexpected by him.'

The Algerian war had made evident to the French military authorities the immense superiority of an infantry trained in this long-continued running. Since 1853, the question was debated whether this system should not be applied to the whole army. General de Lourmel (killed before Sebastopol, 5th November, 1855) had specially drawn the attention of Louis Napoleon to it. Soon after the Crimean war, the *pas gymnastique* was introduced in all French infantry regiments. The time, indeed, is slower, and probably the step, too, shorter, than with the Chasseurs; besides, the long-continued runs of the Chasseurs are much reduced in the line. This was a necessity; the unequal bodily strength and size of the line made the capabilities of the weaker and smaller men the standard of the performance of the whole. But, still, the old sluggish rate of marching can now be overcome at an emergency; a mile or so may now and then be trotted, and, especially, the aptitude of the men to go through their evolutions at the double, admits of that charge, at a run, for some six or eight hundred yards, which carried the French, last year, in a few instants, over those very distances at which the excellent Austrian rifles were most dangerous. The *pas gymnastique* has done a great deal towards the winning of Palestro, Magenta, and Solferino. The run itself gives a vigorous moral impulse to the men; a battalion charging might hesitate when marching at quick time, but the same battalion, trained so as not to arrive out of breath, will, in most cases, go on fearlessly, will arrive comparatively unscathed, and will certainly make a far greater moral impression on a standing enemy, if it charges at a run.

The extreme perfection of the Chasseurs in running may pass for a special arm like theirs, but it would be both impracticable and useless to the mass of the infantry of the line. Nevertheless, the English line, with its better material of men, might easily be made to far surpass the French line in this respect, and like every healthy exercise, this would have a capital effect on the men, bodily and morally. An infantry which cannot alternately run a mile and walk a mile for a couple of hours, will soon be considered slow. As to the volunteers, the great difference of age and bodily strength existing in their ranks, would make it difficult to obtain even this result, but there can be no doubt that gradual training for the double, at distances from half a mile to a mile, would hurt nobody's health and improve wonderfully their efficiency for the field.

22. The French Light Infantry—III[1]

Nothing is neglected in France to develop the physical, mental,

[[1] *Volunteer Journal*, Vol. I, No. 7, pp. 98–9, October 20, 1860. Article by the author of 'A German Account of the Newton Review'.]

and moral powers of every individual recruit, and especially of every Chasseur, in such a manner as to form him into as perfect a soldier as possible. Everything is attended to that can make him strong, active, and nimble, that can give him a rapid glance for advantages of ground, or quickness of decision in difficult situations; everything that will heighten his confidence in himself, his comrades, his arms. Drill, therefore, is but a small portion of a soldier's duties in France; and to our notions, a French battalion on the drill-ground marches, wheels, and does the manual in a shockingly loose manner. But this appears to be a consequence of the national character, and has not, so far, been attended with any bad results. English or German troops seem, themselves, to prefer a stricter system of drill; they obey the command more instantaneously, and, after a certain amount of drilling, will always exhibit more precision in all their movements than the French will ever attain. For the remainder, the system of tactical movements for the drill-ground is nearly the same in France as in England, though it is vastly different on a field of battle.

One of the chief occupations of the French soldier is gymnastic exercise. There is a central military gymnasium in Paris, which forms the teachers for the whole army. There are fifteen to twenty officers from different regiments, and besides, one sergeant from every regiment of the line or battalion of chasseurs, who remain for six months and then relieved by others. The course of exercises gone through is not very different from what is practised in other countries; there appears to be only one original exercise, the escalading of walls, either by putting hands and feet in holes produced by cannon-balls, or by a pole leaned against the wall, or else by means of a rope with a hook thrown over it. This kind of exercise is undoubtedly of practical value, and will contribute a great deal to make the men rely on the use of their hands and feet. The bayonet exercise is also taught in this school; but it is confined to the practising of the various points and guards; the men are never made to actually defend themselves one against the other or against cavalry.

Every garrison, in France, has the necessary conveniences for gymnastic exercise. There is, first of all, a piece of ground set apart for the more common gymnastics, with all the necessary appliances; to this the whole of the soldiers are marched in turns, and have to go through a regular course of instruction as part of their duty. The introduction of this kind of exercise is not yet very old, and is entirely imitated from the Chasseurs, who were the first to be put to gymnastics; after the system had answered so well with them, it was extended to the whole army.

There is, besides, in every barracks, a fencing-room and a dancing-room. In the first, fencing with the small-sword and broad-sword is taught; in the other, dancing and wrestling, which the French call '*la boxe*'. Every soldier may choose which of these exercises he will be taught, but one of them he must learn. Dancing and the small-sword are generally preferred. Single-stick is also taught now and then.

All these exercises, as well as gymnastics, properly so called, are not taught because they are considered necessary in themselves; they are practised because they develop the bodily strength and agility of the soldier generally, and give him greater self-confidence. The fencing and dancing rooms, so far from being the scenes where tedious duty is performed, are, on the contrary, an attraction, tending to keep the soldier in the barracks even in his leisure hours; he will go there for amusement; if, in the ranks, he was but a machine, here, sword in hand, he is an independent man, trying his individual skill against his comrades; and whatever confidence in his own quickness and agility he gains here, it is so much gain for outpost and skirmishing duty, where he is, also, more or less reduced to his own resources.

The new system of skirmishing adopted by the Chasseurs has not only been adopted in the whole French army since, but it has also served as a model for many European armies, among others, for the improved practice adopted in the British army during and after the Crimean war. We shall, therefore, notice but a few of the principal traits, especially as in an engagement the French very often act quite differently, partly in accordance with general orders (as in 1859, in Italy), partly because every latitude is left to officers to act entirely according to circumstances, and partly because all drill regulations must undergo considerable alterations in battle. The skirmishers act in groups of four, each group deploying into one single line, with one man at every five paces), and at most forty paces from group to group. The non-commissioned officers take up a position ten paces behind their sections; the officers, each attended by a guard of four men and a bugler, twenty to thirty paces to the rear. If only part of a company is extended, the captain takes his station half-way between the skirmishers and the support. Taking advantage of cover is the principal thing to be attended to; the dressing of the line as well as the exactness of the intervals are sacrificed to it. The whole line of skirmishers is directed by the bugle alone, the signals numbering twenty-two; besides which, each Chasseur battalion, and every company in it, has a distinctive signal of its own, which is made to precede the signal of command. The officers carry a whistle, which they are, however, to use in extreme cases only; it gives five signals—Caution! Advance! Halt! Retire! Rally!—and is the original of the whistle which some rifle volunteers have adopted as part and parcel of every man's accoutrements, thus depriving their officers of the use of the whistle when it might be necessary. The skirmishers rally by groups of four, if attacked by skirmishing cavalry; by sections and sub-divisions, in irregular compact masses; on the support, where they form a kind of company square; or on the battalion, in case the latter is to act in line or to form square. These various forms of rallying are practised very much, and the French excel in them; and their variety does not create any confusion, as the men are instructed to get rallied any way they can in case of imminent danger, and then to profit of favourable movements to join the larger body to which the signal had called them. The squares are sometimes two, sometimes four deep.

Compared to the old-fashioned system, as adopted in almost all armies before the Chasseurs were organized, this new method had an immense superiority. But it is not to be forgotten that it is, after all, nothing but a set of drill-ground regulations. There is no room in it, as far as it goes, for the intelligence of the individual soldier, and if it was practised on a level plain, it would be compatible with as great pedantry as might satisfy the stiffest martinet. The lines are formed with regular intervals,—they advance, retire, change front and direction same as any battalion in line, and the men are moved by the bugle as so many puppets by a wire. The real practice ground for skirmishers is before the enemy, and here the French had a splendid school for their light infantry in the fearfully broken ground of Algeria, defended by the Kabyles, the bravest, most tenacious, and most wary skirmishers the world ever saw. Here it was that the French developed to the highest degree that instinct for extended fighting and taking advantage of cover which they have shown in every war since 1792; and here the Zouaves especially turned to the best account the lessons given to them by the natives, and served as models to the whole army. Generally a chain of skirmishers is supposed to advance in something like a deployed line, crowding together, perhaps, on points offering good cover, and thinning where they have to pass open ground; occupying the enemy's skirmishers in front, only now and then taking advantage of a hedge or so to put in a little flank fire, and, withal, not expected nor even attempting to do much besides occupying their opponents. Not so the Zouaves. With them, extended order means the independent action, subordinate to a common object, of small groups; the attempt at seizing advantages a soon as they offer; the chance of getting near the enemy's masses, and disturbing them by a well-sustained fire; and, in small engagements, the possibility of deciding them without calling in the masses at all. With them, surprise and ambush are the very essence of skirmishing. They do not use cover merely to open fire from a comparatively sheltered position; they principally use it to creep, unseen, close up to the enemy's skirmishers, jump up suddenly, and drive them away in disorder; they use it to get on the flanks of their opponents, and there to appear unexpectedly in a thick swarm, cutting off part of their line, or to form an ambush, into which they entice the hostile skirmishers, if following too quick upon their simulated retreat. In decisive actions, such artifices will be applicable in the many pauses occurring between the great efforts to bring on decision; but in petty warfare, in the war of detachments and outposts, in collecting information respecting the enemy, or securing the rest of their own army, such qualities are of the highest importance. What the Zouaves are one example will show. In outpost duty, in all armies, the rule is that, especially during the night, the sentries must not sit, nor much less lay down, and are to fire as soon as the enemy approaches, in order to alarm the pickets. Now read the Duke of Aumale's description of a camp of Zouaves (*Revue des Deux Mondes*, 15th March, 1855):

'At night, even the solitary Zouave placed on the brow of yonder hill, and overlooking the plain beyond, has been drawn in. You see no *vedettes*; but wait till the officer goes his rounds, and you will find him speak to a Zouave who is lying flat on the ground just behind the brow, and watchful of everything; you see yonder group of bushes; I should not be at all surprised if on examination you were to find there esconced a few couples of Zouaves; in case a Bedouin should creep up into these bushes in order to espy what is going on in the camp, they will not fire, but despatch him quietly with the bayonet, in order not to shut the trap.'

What are soldiers who have learnt their out-post duty in peace garrisons only, and who cannot be trusted to keep awake except standing or walking, to men trained in a war of ruse and stratagem, against Bedouins and Kabyles? And with all these deviations from the prescribed system, the Zouaves have been surprised only once by their wary enemies.

England has, in the north-west frontier of India, a district very similar, in its military features, to Algeria. The climate is nearly the same, so is the nature of the ground, and so is the border population. Frequent forays and hostile encounters do occur there; and that district has formed some of the best men in the British service. But that these long and highly instructive encounters should not have had any lasting influence upon the mode in which all kinds of light service are carried on in the British army—that after twenty and more years of fighting with Afghans and Baluchis, that part of the service should have been found so defective that French examples had to be hurriedly imitated in order to bring the infantry, in this respect, into a state of efficiency;—this is, certainly, strange.

The French Chasseurs have introduced into the French army: 1. The new system of dress and accoutrements: the tunic, the light shako, the waist belts, instead of the cross belts. 2. The rifle, and the science of its use: the modern school of musketry. 3. The prolonged application of the double, and its use in evolutions. 4. The bayonet exercise. 5. Gymnastics; and, 6. Together with the Zouaves, the modern system of skirmishing. And if we will be sincere, for how much of all this, so far as it exists in the British army, are we not indebted to the French?

There is still plenty of room for improvements. Why should not the British army come in for its share? Why should not the north-western frontier of India, even now, form the troops employed there into a corps capable of doing that for the English army which the Chasseurs and Zouaves have done for the French?

23. French Armaments[1]

According to the *Almanach de Gotha*, which is as good an authority on the subject as can be found anywhere, the war footing of the French army for 1860–1 has been fixed as follows:

1.	Infantry: Guards—12 battalions of Grenadiers, 16 ditto of Voltigeurs, 2 of Zouaves, 1 of Chasseurs; in all 31 battalions. Line—103 regiments of 4 battalions, in all 412 battalions; 3 regiments of Zouaves, 2 of the Foreign Legion, 3 of Turcos (or native Algerian rifles), at 3 battalions each, 24 battalions; Chasseurs, 20 battalions; Zephyrs, or light African (disciplinary) battalions, 3; Pompiers (firemen) of Paris, 1 battalion. In all 491 battalions; or in time of war . . .	515,037 men.
2.	Cavalry: 6 regiments, or 37 squadrons of the Guard; 58 regiments, or 358 squadrons, of the Line; in all 395 squadrons .	100,221 men.
3.	Artillery: 22 regiments—227 batteries (of which 146 are batteries of 6 guns—876 guns are field artillery) . . .	66,007
4.	Engineers	15,443
5.	Train: Sanitary troops, commissariat . . .	24,561
6.	Gendarmes	24,172
7.	Staffs, invalids, military schools, etc . . .	17,324
	Total	762,765

This is the war footing. The peace establishment is as follows:

Infantry	255,248
Cavalry	61,023
Artillery	39,023
Engineers	7,467
Train, etc.	11,489
Gendarmes, invalids, etc. .	41,496
	415,746 men

In January, 1859, a short time before the Italian war broke out, the *Constitutionnel* published an official status of the French army, showing a war establishment of 568,000 men, with a peace establishment of 433,000. How, then, has it been possible within two years to augment the war footing by 200,000 men, while the peace footing has been actually reduced?

[[1] *Volunteer Journal*, Vol. I, No. 22, pp. 283–4, February 2, 1861. Article by the author of the 'History of the Rifle'.]

Again, the annual contingent of able-bodied young men disposable for the army is about 160,000. Of these, under Louis Philippe, between 40,000 and 60,000 were actually enrolled, and found sufficient to keep the army up, in spite of the losses in Algeria. Later on, 80,000, and even 100,000 and more, have been enrolled; the Empire which in peace consumed twice the amount of food for powder than the constitutional monarchy or the republic had required. The time of service is seven years; but, even supposing that of late 100,000 men had been enrolled annually (which is above the average), this would, for seven years, give 700,000 men only; and deducting from these the losses during campaigns and from other causes, there would be scarcely as many as 600,000 men. How, then, are the remaining 163,000 found?

The answer to these two questions is comprised in the late acts of the French Emperor. Before the Italian war, the regiments, hitherto formed in three battalions of eight companies each, are formed in four battalions of six companies each; thus, by merely changing the distribution of the 24 companies of a regiment, four battalions are got instead of three. The size of a battalion has a maximum; above 1,000 men it becomes too strong for one man to command it with his voice, and too unwieldy for quick manœuvring. But the size of a company is far more variable; whether 100 or 250 men, is a matter of choice, not of necessity. By forming the fourth battalions in the way indicated, with the same number of officers and sergeants, the regiment was enabled to muster 4,000 instead of 3,000 strong, as soon as the men were found. During the war, the regiments went out in the strength of three fighting battalions, the fourth forming the depot. Thus, in the fourth battalions of the 100 regiments of the line, the means were found to place 100,000 men more than the old cadres could employ. After the war, the fourth battalions were dissolved, but they have been reinstated again a short time ago. Three more infantry regiments (101st, 102nd, 103rd) have been formed, offering room for 17,000 men more. These new formations account for 112,000 men; and the 51,000 men which remain to be accounted for may constitute the figure to which the army in January, 1859, in consequence of previous losses, was short of its full war complement. This would show that there are cadres now, in the French infantry alone, sufficient to organize the enormous number of men stated above, without any recourse to new formations. But where are the men to be found who are to fill up these cadres?

The regular enrolments of the last seven years will have left on the rolls from 550,000 to 600,000 men. The annual contingent available is about 160,000 men. One year's levy would leave but 50,000 men short, in the worst case; and in case of need, there are the young men who, during the last six years have been entirely liberated from service by drawing favourable lots at the conscription. These might be made available to the tune of some 300,000 at least, but as long habit has made such men consider themselves freed for ever from the obligation to serve, as they are partly married, partly scattered all over the country and hard to find, such a measure would be both unpopular and difficult to carry out.

How, then, does Louis Napoleon make up for the deficiency? By introducing a modification of the Prussian reserve system. Of the 160,000 men available every year, a portion, say one-half, is taken to fill up the vacancies of the standing army. The remainder is enrolled on the reserve list; they are embodied and drilled, the first year two months, the second and third years one month each; they remain liable to be called out for seven years in all, same as the line. Now, we have some reason to believe that if the military surgeons are not over strict in passing the men, and in time of war they get often exceedingly lenient, the annual contingent of 160,000 able-bodied men might, by a stretch, be raised to 200,000; but that we will for the present leave out of the question. In seven years, 160,000 men annually would give an army of 1,112,000 men, and deducting a good round number for losses, there would be fully one million of soldiers. Thus we see that by the new reserve system lately introduced, Louis Napoleon's troops will in a couple of years outgrow the organized bodies ready to receive them. That eventuality, however, is also provided for. In future the four battalions of a regiment are all to be fighting battalions; a fifth battalion is now forming under the name of battalion of instruction, and under the pretext of drilling the men put on the reserve list. This new organization finds room for 103,000 men more, raising the number of men which can be usefully employed by existing corps or cadres to 863,000 men.

Not satisfied with this, the French Government propose to form one more regiment of guards and 17 of infantry of the line; these 18 regiments represent 90 more battalions, or 90,000 men.

Thus, before this year is out, we are sure, from what is known even now, that the French army will be so organized as to be able to stow away comfortably in its battalions, squadrons, and batteries, not less than 953,000 men. And as to finding the men to fill up these organizations, we have seen that up to 700,000 men can be found even this year, without falling back upon men liberated in former years; but, if the universal liability to service, either in the line or reserve, be once acknowledged, it will be easy enough to apply the same principle to the men liberated in the last six years (Napoleon has done the same over and over again in his time), and then there can be no doubt that the full 953,000 men will be soon together.

Here, then, we have the man who unintentionally caused the volunteer movement, responding to it by quietly and noiselessly organizing an army of a million of men, and at the same time laying down twenty iron-cased frigates on the stocks, maybe to escort a fraction of that army across the Channel.

24. General Waldersee on the French Army—I[1]

A short time ago there was published in Berlin, a book on *The French Army on the Drill-ground and in the Field*, which created a great sensation, and rapidly passed through several editions. Although the author merely calls himself 'an old officer' it is no mystery that the book is written by General Count Waldersee, late Minister of War in Prussia. He is a man of very high standing in the Prussian army, where he has particularly distinguished himself by revolutionizing the old pedantic system of teaching the soldier skirmishing, patrolling, outpost, and light infantry duty generally. His new method, to which we may revert on some other occasion, is now introduced in that army. It is remarkable for doing away with all pedantry of forms, and exclusively appealing to the intellectual resources of the soldier in the performance of a duty which can only be carried out well by the intelligent and harmonious co-operation of a number of men. An officer who lays so much stress on the intellectual training of

[[1] *Volunteer Journal*, Vol. II, No. 42, pp. 141–2, June 22, 1861.]

every individual soldier, very naturally took great interest, at all times, in the French army, as the one which is most famous for the individual military intelligence of its men; and we need not, therefore, be astonished if we find that he has made that army the especial object of his studies, and that he has many friends and acquaintances in its ranks, from whom he can obtain valuable information. After the successes of the French against one of the best and bravest European armies, in the Italian campaign of 1859, it became a question of European interest to what circumstances such extraordinary and unvaried victories were owing; and in the above publication General Waldersee gives what he considers to be an elucidation of the subject.

The following is taken from an account of the general character of the French army:

'It partakes of all the good qualities, but also of all the faults and weaknesses, of the French character. Animated by a genuine warlike spirit, it is full of combativeness, thirst for action and for glory, brave and plucky, as it has shown at all times, and more recently on the battlefields of Algeria, the Crimea, and Italy. Everywhere there have been occasions on which both officers and soldiers—particularly among the picked troops—have performed wonders of bravery; and the performances of the French soldiers generally, in these campaigns, are worthy of the highest respect.

'Of great bodily and mental mobility—which, however, is often enough increased to a continuous restlessness—the French soldier is indefatigable and persevering in battle, as well as in hard work of all kinds.

'Self-confident in the highest degree, full of ambition and vanity, every individual soldier has but one desire—to march upon the enemy. He knows no difficulties, he goes by the old French proverb, "If the thing is possible, it is as good as done; if it is impossible, it will be done somehow." Without much reflection—often, indeed, very inconsiderately—he advances, convinced there are no difficulties he cannot overcome. Thus, with the dash and impetuosity inherent to his nation, he always presses for the attack, in which is his chief strength. Besides this, the French soldier is intelligent, handy, particularly adapted for individual fighting, and accustomed to act on his own responsibility. He is inventive and clever in embarrassing situations; he has a peculiar knack of making himself comfortable in a bivouac; of improving bridges, etc., under fire; of putting, at a moment's notice, houses and villages into a defensible state, and of defending them afterwards with the greatest tenacity.

'War is the life-element of an army. The French Government very wisely consider war as the normal state of the troops, and, therefore, at all times and under all circumstances treat them with the same strictness and severity as if actually on a campaign. The regiments are concentrated in camp as frequently as possible, and besides are made to change garrisons constantly, so as not to allow any peace habits to grow up among them. In the same spirit, the drill of the men is exclusively adapted for the purposes of war, and nothing whatever is done for purposes of parade. No corps is ever judged from its style of marching past, and it is, therefore, rather surprising to foreign officers to see French battalions march past—even before the Emperor—with a slovenly gait, in undulating front lines, the men stepping with different feet, and marching at ease with sloped arms.

'But the picture has its dark as well as its bright side. All these good military qualities which urge on the French soldier to advance impetuously, show their brilliant effects *only so long as you allow him to advance*. The *sentiment individuel*, which is at the root of all his qualifications for attack, has its great disadvantages too. The soldier, being principally busied with himself, goes along with the mass as long as it advances successfully; but if this mass be forcibly, and, perhaps, unexpectedly, made to retire, its cohesion, the connection of every individual with his comrade, is soon severed, and the more so as, in such a case, the careless tactical training of the troops—of which, more hereafter—renders all steadiness impossible, and leads to confusion and utter dissolution.

'Add to this that the French are naturally given to envy, and, with all their national levity in critical moments, are apt to be suspicious of others. The French soldier follows his officers eagerly and willingly into battle, but only so long as these officers are in front of him, and literally lead him on. This is what the soldiers expect, and when advancing under fire they express it by shouting, "Epaulettes to the front!" Thus field officers and generals have generally to march to the charge in front of their troops—the very place, certainly, for a general—and this explains the excessive losses the French always have had in officers. But if a retreat becomes inevitable, confidence in the officers will soon disappear, and, in extreme cases, make room for open disobedience. From these causes, a retreat, energetically forced upon a French army, has always been disastrous to it, and will ever be so.'

General Waldersee might have added a great deal more on the facility with which the confidence of the French soldier in his officers melts away under adverse circumstances. The confidence of the men in their immediate superiors, even after repeated unsuccess, is the best standard of discipline. Measured by this, the French are not much better than totally undisciplined levies. It is a matter of course for them that they never can be beaten except by 'treachery'; and whenever they lost a battle and had to retreat more than a few hundred yards—whenever the enemy surprised them by an unexpected move, they regularly raised the cry, 'We are betrayed!' So much is this part and parcel of the national character, that Napoleon, in his memoirs (written long after the fact, at St. Helena), could impute, by insinuation, some kind of treacherous action to most of his generals; and that French historians—military and otherwise—could amplify these insinuations into the most wonderful romances. As the nation of the generals, so does the soldier think of his regimental and company officers. A few hard knocks, and discipline is completely at an end; and thus it is that, of all armies, the French have made the most disastrous retreats.

25. General Waldersee on the French Army—II[1]

Of the mode of recruiting the soldiers and officers, Waldersee gives the following account:

'The French soldier is recruited by drawing lots among the young men of the country; but every man has the right of paying a sum fixed by Government for a substitute. This sum flows into a fund, administered by the Government, from which the substitute receives a small sum as bounty on enlisting, and the

[1 *Volunteer Journal*, Vol. II, No. 44 ,p. 159, July 6, 1861.]

remainder on the expiration of his term, the interest being paid to him during his time of service. The sum owing to him may, however, be partially or totally forfeited by crime or bad conduct. Thus the Government have the selection of substitutes entirely in their own hands, and are in the habit of enlisting, as much as possible, men only who have already served one term of seven years, and who have proved themselves reliable and well conducted. A great many old soldiers are thus secured to the army, and from them most of the non-commissioned officers are selected. The term of service is seven years; of this time, however, the greater portion of the men are but four or five years actually with the colours, spending the remainder on furlough.

'The non-commissioned officers are selected with great care and tested with great regard by the officers. They are mostly distinguished, not only by an excellent character and a perfect knowledge of the details of their duty, but also by intelligence, independence, a fine soldier-like bearing, and a certain dignity, especially in their relations with the privates, over whom they know very well how to maintain the great authority which the regulations have given them. As every non-commissioned officer is eligible for a commission, they manage to keep the privates at a respectful distance, while, on the other hand, they use every effort to distinguish themselves and give a good example to their subordinates.

'At present the majority of the non-commissioned officers consist of substitutes. A few only are made corporals and sergeants during their first term of service, and among them, particularly those young men who, having had a good education, and finding themselves excluded by the great throng of candidates from the military schools, enlist voluntarily in the army in order to try for a commission. Such young men very soon advance to the position of non-commissioned officers, and on passing the practical military examination prescribed for sergeants before they can be made sub-lieutenants, very often receive a commission after having served from two to four years.

'The generality of officers promoted from the ranks receive their commissions after from 9 to 12, and often after from 15 to 20 years only. Of 170 such officers, taken at random, 16 received commissions after from 2 to 4, 62 after 5 to 8, 62 after 9 to 12, and 30 after from 13 to 20 years' service. The first 16 belonged to the class of educated young men; the 62 who received commissions after from 5 to 8 years, were promoted for distinction before the enemy. Thus, in time of peace, promotion from the ranks, even in France, is slow work.

'The officers recruit themselves partly from the ranks, as stated above, and partly (in times of peace principally), from the military schools, where the young men have to attend for two years, after which, on passing a severe examination, they at once receive commissions. These two classes of officers keep at a great distance from each other; the pupils of the military colleges and the educated men promoted from the ranks, looking down with disrespect upon the old sub-lieutenants and lieutenants who gained their epaulettes by long service; the officers, even of the same battalion, form anything but that compact body which they do in almost every other army. Yet those men who were raised from the comparatively less educated portion of the ranks (and who now, after the heavy losses in the Crimea and Italy, form the greater portion of the subalterns), are very useful in their way. Though very often positively ignorant, and sometimes rough, and scarcely above the sergeant in character or manners, they are generally clever within their sphere of action, perfectly at home in their duty, conscientious, strict, and punctual; they know exceedingly well how to treat the soldier, how to take care of him, how to stimulate him by their example, both in garrison life and under fire. Beside this, they at present mostly possess a good deal of experience in camp life, marching and fighting.

'On the whole, the French officer is intelligent and eager for war; he knows what he is about, and especially under fire—he knows how to act on his own responsibility, and how to excite the men by the example of his own bravery. Add to this—for the majority of them—a good deal of campaigning and fighting experience, and we must say that they are possessed of qualities which place them very high in their profession.

'Promotion is given either by seniority or by selection. In peace, two by seniority to one by selection; in war, the reverse. But selection is generally limited to the educated class of officers, while the mass of those raised from the ranks are promoted by seniority only, and thus attain their captaincy at a rather advanced age. This is about the highest step they ever reach, and they are generally quite satisfied to be able to retire on a captain's pension.

'Thus it happens that in the French army you see a good many subalterns of from 30 to 40, and a good many captains approaching 50; while among field-officers and generals there are a great many comparatively young men. This is no doubt a great advantage; and the continued wars in Africa, the Crimea, and Italy, having considerably quickened promotion, have brought still more young men into high commands.

'To show the proportion in which promotion to the higher grades is dealt out to the two classes of officers, the following statement of officers killed and wounded, or employed in high commands in Italy, will be read with interest: From the military schools; 34 generals, 25 colonels commanding regiments, 28 other field-officers, 24 captains, 33 lieutenants and sub-lieutenants. From the ranks: 3 generals, no colonels commanding, 8 field-officers, 66 captains, 95 subalterns.

'The generals proceed less from the staff and the scientific or select corps than from the generality of the field-officers. They therefore are mostly wanting in military instruction of a higher order; a few among them only have *les vues larges*. Badly up in strategy, they are rather clumsy in handling large bodies of troops, and therefore much in want of superior orders or scientific assistance; so that very often in the field, as on the drill ground, they receive a regular programme of the movements to be gone through for engaging an action. On the other hand, they are full of common sense, and ready at inventing expedients; they know the practical part of their duty, are zealous, ambitious, and devoted to the service. Their habit of acting independently gives them the necessary vigour under fire. They know no difficulties, act at once on every emergency, without awaiting or sending for orders; are not afraid of responsibility; and, brave like every Frenchman, they always personally lead on their troops.

'Most of them have fought in Algeria, the Crimea, and Italy, and, therefore, are in possession of a valuable store of warlike experience. Of the generals engaged in Italy in 1859 there were twenty-eight old Africans, eighteen of whom had also fought in the Crimea. One general alone (Partouneau) made his first campaign in Italy.

'This continued fighting has endowed the French army with

a younger body of generals than any other army can boast of. To keep this up in time of peace, lieutenant-generals retire on half-pay at 65, and major-generals at 60 years of age.

'In short, the French generals must be regarded as comparatively young and bodily active, intelligent, energetic, experienced in war and well adapted for it, though but a few have, so far, shown themselves unusually clever and well acquainted with the handling of large bodies of troops, and though neither the Crimean nor the Italian war have developed any extraordinary military genius.'

26. Waldersee on the French Army—III[1]

Passing to the drill-practice of the French, our author says:

'The recruit, boorish and clumsy as he is when joining his regiment, nevertheless often enough, before a fortnight is over, and before even he may have received his full equipment, stands sentry with the dignity and authority of an old trooper, and very soon becomes formed by the careful *individual training* which he is made to go through. Though company and battalion drill leave very much to be desired, every individual soldier is carefully trained to gymnastic and bayonet exercise, fencing with the small sword, and long running at double-quick time.

'On the drill-ground the infantry is generally without steadiness, loose, and therefore rather slow; but on a march it is exceedingly quick, and broken to long marches, great portions of which are made at the double; which pace is very often used in action, and to no mean advantage. These are the performances by which the excellence of a body of troops is judged in France; it is never judged by its drill, much less by mere marching past. The fact is, the French cannot march past in good order, because they are defective in that drill in detail which, after all, is necessary to every good body of troops.'

Talking about drill, our author gives the following anecdote of Napoleon I:

'Napoleon was well aware of the drawbacks inherent to this loose system of drill and did his best to redress it. Under his iron rod, precision of drill was adhered to as much as it was

[[1] *Volunteer Journal*, Vol. II, No. 46, pp. 174–5, July 20, 1861.]

possible with Frenchmen—though he himself was no very good drill-master. One day, at Schönbrunn, in 1809, he had the idea of drilling himself a battalion of his guards; to make them *faire la théorie*, as the French call it. He drew his sword, and gave the word; but after having ordered a few movements, he got his men into such utter confusion that he called out, putting his sword back into the scabbard, "The devil take your —— theory! Set that mess right again." (*Que le diable emporte votre f—— théorie! Redressez cette cochonnerie!*)'

About the 'Turcos', the native Algerian troops, we find the following remarkable statement:

'According to reports received from French officers, the Turcos above all disliked an engagement with the Austrian Rifles. Whenever they met them, they not only refused to advance, but threw themselves down, and, like the camels of the desert, could not be induced either by threats or by blows, to rise to the attack.'

On the drill-ground of an infantry regiment:

'Recruit drill is gone through in a very pedantic manner, but still very superficially; little attention is paid to the bearing of the individual men, and thus, the regulations are carried out (in company and battalion-drill) in a positively slovenly manner. Very little care is taken that the men stand properly at attention, that the dressing is good, the line well closed up, or even that the men step out with the same foot. It appears to be sufficient that the men be *there*, and arrive together, somehow or other. An army accustomed to such a loose system of drill will certainly not show to any great extent the disadvantages it entails, so long as it continues to advance. Still, this system must exercise a very bad influence on discipline and order in action; and whenever a retreat under fire becomes inevitable, it may bring on the most serious consequences. This is the reason why the attempt at a retreat in good order has so often proved dangerous to the French, and why a retreat forced upon them by a solid, well-schooled army will always prove disastrous to them.'

After disposing of the drill, General Waldersee gives an epitome of Marshal Bugeaud's principles of fighting (the same which we have in great part translated in preceding numbers of the *Volunteer Journal* under the heading: 'On the moral elements in fighting'). With these principles he fully coincides, attempting at the same time to prove—and not without success—that most of them are old practical rules, to be found already in the instructions of Frederick the Great. We pass over this, as well as over a lengthy strategical criticism of the campaign in Italy in 1859 (in which not less than eighteen distinct blunders of General Gyulai are shown up), in order to come to the observations on the mode of fighting of the French in that campaign.

The most essential principles of this method are:

1. To act on the offensive whenever this is in any way possible.

2. To treat protracted firing with contempt, and to pass as soon as ever possible to a charge with the bayonet, at the double.

This being once known, it has been very generally concluded that the French always and everywhere, with a complete disregard of all tactical forms, had rushed upon the Austrians, and that they had always instantly, and without further ado, run them down or driven them away.

But the history of the campaign is there to prove that this was far from being the case. On the contrary, it shows:

1. That the French certainly did in most cases, not always, attack their opponents impetuously in double quick time, but that scarcely ever did they conquer them at the first charge. Not only were they generally unsuccessful in this, but in most cases they were defeated with loss in several repeated attacks, so that during action they retreated nearly as often as they advanced.

2. That often enough they charged without firing, but, once repelled, they were obliged to carry on the engagement by firing, which firing lasted for some time, though interrupted by repeated bayonet charges. At Magenta and Solferino such firing engagements lasted several hours.

The author now gives, from reports received both from French and Austrian officers, an account of the tactical formations applied by the French during the Italian campaign, by extracts from which we shall conclude this article.

27. Waldersee on the French Army—IV[1]

Our author, after describing the general character and principles of fighting of the French army, proceeds to give an account of

[[1] *Volunteer Journal*, Vol. III, No. 62, p. 299, November 8, 1861.]

the tactical formations employed by them in the Italian campaign of 1859:

'A French army division is composed of two brigades, the first of which has a battalion of chasseurs, and two regiments (of three battalions each) of the line, while the second has only two regiments (or six battalions) of the line. Each battalion has six companies.

'In the line of battle, the first brigade forms the first line, the battalions being formed in columns at half distance with full deploying intervals between them, and covered by a line of skirmishers. The second brigade stands in second line, 250 yards to the rear, the battalions equally in columns at half distance, but with only half deploying intervals between them; they are generally placed behind one of the *wings* of the first line.

'The formation of column generally adopted in the Italian war was what the French call column of divisions—two companies with them being called a division. The six companies are ranged two in front, two at half-distance behind them, and again two companies at half-distance behind the second pair of companies. This column may either be formed on the two centre companies or on the two extreme companies of either wing. With the Guards, who were all picked men, it was always formed on the two centre companies, and thereby (same as in the English double column on the two centre sub-divisions) the time both for forming column and for deployment was abridged by one-half; but with the line it was generally formed on the two right companies. The reason was, that by this method the "grenadier" company (No. 1) was placed in the front of the column, while the light or "voltigeur" company (No. 6) came to the rear. Thus these two companies, consisting of picked men, formed, so to speak, a framework in which the less reliable four "centre companies" were encompassed; and, moreover, in case the two rear companies were ordered to extend as skirmishers, the light company was one of them, while the grenadier company, in the front line, remained together unless the whole battalion had to extend.

'For an army fighting chiefly, not in line, but by a combination of skirmishes and columns, this formation offers great advantages. One-third of the men (the two front companies) are always in a position to make use of their fire-arms, while at the same time deployment is simple and can be got through very quickly. The great distance between the component parts of the column (half company distance or about 40 yards) tends very much to reduce the ravages which artillery makes in closer columns; and when it is borne in mind that, as a rule, two companies were extended, so that the whole column consisted of two companies in front, and two at 40 yards behind them, it is seen that this formation approaches the line as much as possible; the two rear companies acting rather as a reserve or second line to the two front ones than as that bodily support which is generally supposed to be given by the rear men to the front line in continental columns of attack. Moreover, although deployments into line did now and then occur in the Italian campaign, the ground in Lombardy is such that fighting in line is positively impossible. In these small fields, intersected by hedges, ditches, and stone walls, and covered, besides the corn, with mulberry trees connected one with another by vine branches; in a country where the lanes, running between high walls, are so narrow that two carts can scarcely pass each other—in such a country all regular formation often cease so soon as troops advance to close with the enemy. The only thing necessary is to have plenty of skirmishers in front, and to dash with the compact masses on to the most important points. Now, for such a purpose, there could be no better formation than that selected by the French. One-third of the battalion skirmishing—no supports, the column at 100 yards to the rear being support enough—the whole advancing rapidly, the skirmishers, when near enough, clearing the front of the battalion and hovering on its wings; the first line giving a volley and charging; the second, 40 yards to the rear, following as a reserve and keeping as much order as the ground will allow. We must admit that this method seems very well adapted for all purposes of attack in such ground, and will keep the men as much as possible together, and under the control of their officers.

'Wherever the ground was open enough to admit of regular movements, the attack was carried out in this way—the skirmishers engaged the enemy until the order was given for the column to advance; the supports—if supports there were—forming on the flanks of the line of skirmishers, and extending themselves to the front of either wing, in order to envelop and give cross-fire to an advancing enemy; when the column came up to the line of skirmishers, the latter crowded in the intervals of battalions, advancing in a line with the head of the column; at twenty yards from the enemy the head of the column fired a volley and charged. When the ground was very thickly covered, as many as three or four companies of a battalion were extended, and cases are reported (at Magenta, the Turcos) where whole battalions extended as skirmishers.

'Against an Austrian bayonet attack, a method similar to that prescribed by the British regulations for street firing (battalion drill, section 62) was sometimes employed. The leading companies of the column gave a volley, faced outwards, and filed to the rear, where they re-formed; the succeeding companies did the same, until after the rear companies had fired their volley and cleared the front, the whole battalion charged the enemy.

'In decisive moments, the soldiers were ordered to deposit their knapsacks on the ground, but to provide themselves with some bread and all the ammunition it contained, which they stored away about their persons as best they could. This is the origin of the fable, "That the Zouaves carried their cartridges habitually in their breeches' pockets."

'At Magenta, the Zouaves and the 1st Grenadier Guards deployed for a time, and fired by files and by ranks; at Solferino, too, the division of voltigeurs of the Guards (twelve battalions) deployed in a single line before going into action, but when actually engaged, they seem to have been in the usual column. As both these deployments were made under the immediate command and in the presence of Louis Napoleon, there can scarcely be any doubt that he ordered them from some recollection of English line manœuvres; but in both cases the predilection of the French officers for their own national mode of fighting and the nature of the ground, appear to have prevailed as soon as the real tug of war came on.

'In the attack on a village, several columns, preceded by thick swarms of skirmishers, were launched; the weaker column, destined to attack the front of the position was held back to the last, while stronger columns turned the flanks of the village. The troops who *took* the place at once occupied and fortified it, while *the reserves* pursued the enemy. To defend a village, the French trusted more to the reserves behind it or on its flanks than to a strong garrison in the houses themselves.'

With this abstract of the tactical formations of the French

army of Italy in 1859, we take leave of Count Waldersee's work. Although England is far less rough ground for fighting than Lombardy, still her numerous fences, ditches, clusters of trees, and coppices, combined with the undulating nature of the ground, and the deep wooded ravines cut into it, make her a far rougher battlefield than the large uninterrupted plains of Northern France, Belgium, and Germany. If ever a French army should attempt a descent on English soil, there can be little doubt that the formations of its infantry would be very similar to those employed in Italy; and that is the reason why we think these formations not without interest to English volunteers.[1]

[[1] The editors have been unable to confirm Engels's assertion that Waldersee was the author of the anonymous work from which these extracts were translated.]

PART IV
THE CIVIL WAR IN THE U.S.A.

28. Lessons of the American War[1]

When, a few weeks back, we drew attention to the process of weeding which had become necessary in the American volunteer army, we were far from exhausting the valuable lessons this war is continually giving to the volunteers on this side of the Atlantic. We therefore beg leave again to revert to the subject.

The kind of warfare which is now carried on in America is really without precedent. From the Missouri to Chesapeake Bay, a million of men, nearly equally divided into two hostile camps, have now been facing each other for some six months without coming to a single general action. In Missouri, the two armies advance, retire, give battle, advance, and retire again in turns, without any visible result; even now, after seven months of marching and counter-marching, which must have laid the country waste to a fearful degree, things appear as far from any decision as ever. In Kentucky, after a lengthened period of apparent neutrality, but real preparation, a similar state of things appears to be impending; in Western Virginia, constant minor actions occur without any apparent result; and on the Potomac, where the greatest masses on both sides are concentrated, almost within sight of each other, neither party cares to attack, proving that, as matters stand, even a victory would be of no use at all. And unless circumstances foreign to this state of things cause a great change, this barren system of warfare may be continued for months to come.

How are we to account for this?

The Americans have, on either side, almost nothing but volunteers. The little nucleus of the former United States regular army has either dissolved, or it is too weak to leaven the enormous mass of raw recruits which have accumulated at the seat of war. To shape all these men into soldiers, there are not even

[[1] *Volunteer Journal*, Vol. III, No. 66, pp. 334–5, December 6, 1861.]

drill-sergeants enough. Teaching, consequently, must go on very slow, and there is really no telling how long it may take until the fine material of men collected on both shores of the Potomac will be fit to be moved about in large masses, and to give or accept battle with its combined forces.

But even if the men could be taught their drill in some reasonable time, there are not officers enough to lead them. Not to speak of the company officers—who necessarily cannot be taken from among civilians—there are not officers enough for commanders of battalions, even if every lieutenant and ensign of the regulars were appointed to such a post. A considerable number of civilian colonels are therefore unavoidable; and nobody who knows our own volunteers will think either McClellan or Beauregard over timid if they decline entering upon aggressive action or complicated strategical manœuvres with civilian colonels of six months' standing to execute their orders.

We will suppose, however, that this difficulty was, upon the whole, overcome; that the civilian colonels, with their uniforms, had also acquired the knowledge, experience, and tact required in the performance of their duties—at least, as far as the infantry is concerned. But how will it be for the cavalry? To train a regiment of cavalry, requires more time, and more experience in the training officers, than to get a regiment of infantry into shape. Suppose the men join their corps, all of them, with a sufficient knowledge of horsemanship—that is to say, they can stick on their horses, have command over them, and know how to groom and feed them—this will scarcely shorten the time required for training. Military riding, that control over your horse by which you make him go through all the movements necessary in cavalry evolutions, is a very different thing from the riding commonly practised by civilians. Napoleon's cavalry, which Sir William Napier (*History of the Peninsular War*)[1] considered almost better than the English cavalry of the time, notoriously consisted of the very worst riders that ever graced a saddle; and many of our best cross-country riders found, on entering mounted volunteer corps, that they had a deal to learn yet. We need not be astonished, then, to find that the Americans are very deficient in cavalry, and that what little they have consists of a kind of Cossacks or Indian irregulars (rangers) unfit for a charge in a body.

[[1] The correct title is *History of the War in the Peninsula* (1828–40).]

For artillery, they must be worse off still; and equally so for engineers. Both these are highly scientific arms, and require a long and careful training in both officers and non-commissioned officers, and certainly more training in the men too, than infantry does. Artillery, moreover, is a more complicated arm than even cavalry; you require guns, horses broken in for this kind of driving, and two classes of trained men—gunners and drivers; you require, besides, numerous ammunition-waggons, and large laboratories for the ammunition, forges, workshops, etc.; the whole provided with complicated machinery. The Federals are stated to have, altogether, 600 guns in the field; but how these may be served, we can easily imagine, knowing that it is utterly impossible to turn out 100 complete, well-appointed, and well-served batteries out of nothing in six months.

But suppose, again, that all these difficulties had been overcome, and that the fighting portion of the two hostile sections of Americans was in fair condition for their work, could they move even then? Certainly not. An army must be fed; and a large army in a comparatively thinly-populated country such as Virginia, Kentucky, and Missouri, must be chiefly fed from magazines. Its supply of ammunition has to be replenished; it must be followed by gunsmiths, saddlers, joiners, and other artisans, to keep its fighting tackle in good order. All these requisites shone by their absence in America; they had to be organized out of almost nothing; and we have no evidence whatever to show that even now the commissariat and transport of either army has emerged from babyhood.

America, both North and South, Federal and Confederate, had no military organization, so to speak. The army of the line was totally inadequate, by its numbers, for service against any respectable enemy; the militia was almost non-existent. The former wars of the Union never put the military strength of the country on its mettle; England, between 1812 and 1814, had not many men to spare, and Mexico defended herself chiefly by the merest rabble. The fact is, from her geographical position, America had no enemies who could anywhere attack her with more than 30,000 or 40,000 regulars at the very worst; and to such numbers the immense extent of the country would soon prove a more formidable obstacle than any troops America could bring against them; while her army was sufficient to form a nucleus for some 100,000 volunteers, and to train them in reasonable time. But when a civil war called forth more than a million of fighting men, the whole system broke down, and everything had to be begun at the beginning. The results are before us. Two immense, unwieldy bodies of men, each afraid of the other, and almost as afraid of victory as of defeat, are facing each other, trying at an immense cost to settle down into something like a regular organization. The waste of money, frightful as it is, is quite unavoidable, from thc total absence of that organized groundwork upon which the structure could have been built. With ignorance and inexperience ruling supreme in every department, how could it be otherwise? On the other hand, the return for the outlay, in efficiency and organization, is extremely poor; and could that be otherwise?

The British volunteers may thank their stars that they found, on starting, a numerous, well-disciplined, and experienced army to take them under its wings. Allowing for the prejudices inherent to all trades, that army has received and treated them well. It is to be hoped that neither the volunteers nor the public will ever think that the new service can ever supersede, in any degree, the old one. If there are any such, a glance at the state of the two American volunteer armies ought to prove to them their own ignorance and folly. No army newly formed out of civilians can ever subsist in an efficient state unless it is trained and supported by the immense intellectual and material resources which are deposited in the hands of a proportionately strong regular army, and principally by that organization which forms the chief strength of the regulars. Suppose an invasion to threaten England, and compare what would be then done with what is unavoidably done in America. In England, the War Office, with the assistance of a few more clerks, easily to be found among trained military men, would be up to the transaction of all the additional labour an army of 300,000 volunteers would entail; there are half-pay officers enough to take, say, three or four battalions of volunteers each under their special inspection, and, with some effort, every battalion might be provided with a line-officer as adjutant and one as colonel. Cavalry, of course, could not be improved; but a resolute reoganization of the artillery volunteers—with officers and drivers from the Royal Artillery—would help to man many a field-battery. The civil engineers in the country only wait for an opportunity to receive that training in the military side of their profession which would at once turn them into first-rate engineer officers. The commissariat and transport services are organized, and may soon be made to supply the wants of 400,000 men quite as easily as those of 100,000. Nothing would be disorganized, nothing upset; everywhere there would be aid and assistance for the volunteers, who would nowhere have to grope in the dark; and—barring some of those blunders which England cannot do without when first she plunges into a war—we can see no reason why in six weeks everything should not work pretty smoothly.

Now, look to America, and then say what a regular army is worth to a rising army of volunteers.

29. The War in America[1]

The real opening of the campaign in this war dates from the advance of the Union forces in Kentucky. Not before Missouri and Western Virginia had been finally reconquered did this advance commence. The Secessionist troops held three strong positions—entrenched camps—in the State of Kentucky; Columbus, on the Mississippi, on their left; Bowling Green, in the centre; Mill Spring, on the Cumberland River, on their right. Their line thus extended fully 250 miles as the crow flies. By road, the distance certainly was 300 miles east and west. Such an extended line precluded all possibility of these corps supporting each other, and gave the Federal forces a chance of attacking each of them separately with superior forces. There was no great risk in such a course, as none of the three Secessionist corps were strong enough to advance, even if unopposed, beyond the Ohio River. The great mistake in the Secessionist position was the attempt to occupy everything, and the consequent dissemination of the troops. One strong central entrenched camp, destined to be the prepared battlefield for a decisive action, and held by the main body, would have defended Kentucky far more efficiently; for it must either have attracted the main body of the Federals, or placed them in a disadvantageous position if they attempted to march past it without noticing this strong concentration of troops. As it was, the Federals attempted to attack these three

[[1] *Volunteer Journal*, Vol. IV, No. 80, pp. 9–10, March 14, 1862. Article by the author of *Essays Addressed to Volunteers*.]

camps one after another, and to manœuvre their enemy out of them, so as to compel him to fight in the open. This plan was completely in accordance with the rules of military art, and it was executed with a vigour and rapidity which deserves much commendation, as well as the perfect success obtained. Towards the middle of February, a body of 15,000 Federals moved upon Mill Spring, which was held by about 10,000 Confederates. The Federals manœuvred so as to make their adversaries believe that but a weak force was in the neighbourhood, and the Confederate general, Zollicoffer, at once took the bait thrown out to him. He marched out of his works, attacked the first Federal body he met, but very soon found that he had to do with a force superior to his own in numbers, and at least its equal in spirit and discipline. He fell, and his troops were as completely routed as the Federals had been at Bull's Run. But this time the victory was followed up far differently. The beaten army were pursued very closely until they arrived, broken, demoralized, and deprived of their field artillery and baggage, at their camp of Mill Spring. The camp was constructed on the northern shore of the Cumberland River, so that the troops, in case of another defeat, had no retreat but by a few steamers and boats across the river. We shall find that almost all these Secessionist camps were thus placed on the enemy's side of a river. Such an encampment is perfectly correct, and of the greatest utility—when there is a bridge. The camp, in that case, serves as a bridge-head, and gives to its occupants the chance of throwing their forces at will on either bank of the river, by which alone they obtain a perfect command over it. But to do the same thing when there is no bridge, is to place your troops in a position where they have no retreat after an unlucky engagement, and when, therefore, they will either have to surrender or be massacred and drowned, same as the Federals were whom General Stone's treachery had sent across the Potomac at Ball's Bluff. Accordingly, when the defeated Secessionists reached their camp at Mill Spring, the fact at once became patent to them that unless they could beat off an attack on their entrenchments, they would have to surrender very speedily. After the experience of the morning, they had no longer any confidence in their powers of resistance; and when the Federals, next morning, advanced to attack the entrenched camp, they found that the enemy had taken advantage of the night to cross the river, abandoning camp, baggage, artillery, and stores. Thus the extreme right of the Confederate line was driven back into Tennessee; and Eastern Kentucky, where the population are chiefly Union men, was reconquered for the Union.

About the same time—the second half of January—the preparations for dislodging the Secessionists from Columbus and Bowling Green were commenced. A strong fleet of mortar-boats and iron-clad gunboats had been got ready, and the news was spread everywhere that they were to accompany the march of a strong army down the Mississippi, from Cairo to Memphis and New Orleans. A ridiculously conspicuous reconnaissance was made towards Columbus. The retreat of this strong body of troops, which did not effect anything, even looked like a serious check to the Union troops. But it seems that all these demonstrations on the Mississippi were mere blinds. When everything was ready, the gunboats were quietly removed into the Ohio, and thence into the Tennessee River, which they steamed up to Fort Henry. This place, together with Fort Donelson, on the Cumberland River, formed a second line of defence of the Secessionists in Tennessee. The position was well chosen; for if they had retreated behind the Cumberland River, this would have covered their front, and the Tennessee river their left flank, while the narrow strip of land between the two would have been sufficiently covered by the two camps just named. But the rapid action of the Federals broke through the second line before even the left and centre of the first was attacked.

In the first week of February, the Federal gunboats appeared before Fort Henry, and shelled it with such effect that it at once surrendered. The garrison escaped to Fort Donelson, the land force of the expedition not being strong enough to invest the place. Then the gunboats steamed down the Tennessee again, up the Ohio, and up the Cumberland, towards Fort Donelson; only one gunboat boldly steamed up the Tennessee, right through the heart of the State of Tennessee, skirting the State of Mississippi, and penetrating as far as Florence, in Northern Alabama, where a series of flats and swamps (the so-called mussle (*sic*) shoals) stop further navigation. The single fact of one gunboat performing this long journey (at least 150 miles) and returning without ever being attacked, proves in itself that there must be, along this river at least, a strongly prevailing Union sentiment, which no doubt will tell very powerfully if the Federals should penetrate so far.

The naval expedition up the Cumberland now concerted its movements with those of the land forces under General Halleck and Grant. The Secessionists at Bowling Green were deceived as to the Federal movements, and remained quiet and confident in their camp, while a week after the fall of Fort Henry, Fort Donelson was invested on the land side by 40,000 Federals and menaced on the river by a powerful fleet of gunboats. Same as Mill Spring and Fort Henry, the entrenched camp of Fort Donelson was constructed with its rear to the river and no bridge for a retreat. It was the strongest place the Federals had as yet attacked. The works were not only constructed with much greater care, but, besides, it was large enough to shelter the 20,000 men which held it. On the first day of the attack, the gunboats silenced the fire of the batteries facing the river and shelled the interior of the works, while the land forces drove in the enemy's outposts and compelled the main body to take shelter close under the guns of their works. On the second day, the gunboats, having suffered severely the day before, appear to have done little work, but the land forces had to fight a long and sometimes severe battle with the columns of the garrison, which tried to break through their right in order to keep open the line of retreat towards Nashville. But a vigorous attack of the Federal right upon the Secessionist left, and strong reinforcements sent to the Federal left, decided the victory in favour of the assailants. Several outworks had been stormed; the garrison, hemmed in within their inner lines of defence, without any chances of retreat, and evidently not in a condition to resist an assault next morning, surrendered on the third day unconditionally. General Floyd escaped on the evening of the second day, it is said, with 5,000 men. It is not quite clear how that was possible; the number is too large to have been stowed away on steamers during the night; but still they may have successively crossed the river, and escaped along its right bank. The whole of the artillery, baggage, and stores, together with 13,300 prisoners, fell into the hands of the Unionists; 1,000 more prisoners were made next day, and on the appearance of the Federal advanced guard, Clarksville, a town higher up the river, surrendered with great quantities of stores, collected there for the Secessionist troops.

Whether Nashville has also fallen, appears very uncertain, and we can scarcely believe it. As it is, these successors of the Federals, in the short space of three weeks, are quite enough for them to be satisfied with. Columbus, the only place the Secessionists now hold in Kentucky, they can continue to hold at very great risks only. If they lose a decisive battle in Tennessee, the garrison of Columbus, cannot escape being compelled to surrender, unless the Federals commit very great blunders. And that the Confederates are now compelled to fight a decisive battle in Tennessee, is one of the great results of the Federal victories. They have concentrated, we are told, 65,000 men at and about Nashville; it may be that they have succeeded in collecting even a larger force. But the combined troops of Halleck, Grant, Buell and Thomas, together with the reserve now hurrying up from the camps of instruction in Kentucky, Ohio, Indiana, and Illinois, will enable the Federals to outnumber them; and with their *morale* necessarily much raised above that of their adversaries by the late successes, and with a strong Union party among the population to keep them well informed of the movements of the enemy, we do not see that they have any reason to be afraid of the issue.

PART V

THE SCHLESWIG-HOLSTEIN WAR OF 1864

30. THE STRENGTH OF THE ARMIES IN SCHLESWIG[1]

THERE are most absurd reports afloat as to the relative strength of the contending armies in the Danish war. It is generally supposed that the Danes are outnumbered in the proportion of one Dane to at least three Germans. To show how little this is in accordance with facts, I propose to give a detailed statement of the strength of each army, as far at least as its infantry is concerned; for as to cavalry and artillery it would at present be very difficult to get precise information.

Before the outbreak of hostilities, the Danes had the following troops in Schleswig, viz.:

First Division; Commander, Lieutenant-General Gerlach:	*Battalions*
1st Brigade, 2nd and 22nd Infantry Regiments	4
2nd Brigade, 3rd and 18th Infantry Regiments	4
3rd Brigade, 17th and 19th Infantry Regiments	4
Second Division; Major General Du Plat:	
4th Brigade, 4th and 6th Infantry Regiments	4
5th Brigade, 7th and 12th Infantry Regiments	4
6th Brigade, 5th and 10th Infantry Regiments	4
Third Division; Major General Steinmann:	
7th Brigade, 1st and 11th Infantry Regiments	4
8th Brigade, 9th and 20th Infantry Regiments	4
9th Brigade, 16th and 21st Infantry Regiments	4
Total Battalions	36

Or, at 800 men for a battalion (the full complement is 870 men and officers), say	28,800 men
Cavalry, 4½ regiments, at 560 men	2,500 „
Artillery about	3,000 „
Total Danish forces	34,300 men

Exclusive of several battalions, both of line and reserve, which

[1] Letter signed 'F.E.' in the *Manchester Guardian*, February 16, 1864.

were sent to Schleswig in the first days of February, but as to which it has been impossible to ascertain any particulars.

The Austrians have sent to the seat of war the sixth army corps, consisting of the following troops:

	Battalions
General Gondrecourt's Brigade:	
Infantry Regiment, King of Prussia	3
Ditto, Baron Martini	3
Chasseur Battalion, No. 18	1
General Nostitz's Brigade:	
Infantry Regiment, King of the Belgians	3
Ditto, Grand Duke of Hesse	3
Chasseur Battalion, No. 9	1
General Thomas's Brigade:	
Infantry Regiment, Count Coronini	3
Ditto, Prince Holstein	3
A Chasseur Battalion, number not stated	1
General Dormus's Brigade:	
Two Infantry Regiments and one battalion of Chasseurs, numbers and names not stated	7
Total battalions	28

Or at 800 men per battalion (which is a high estimate for the present organization of the Austrian army)	22,400 men
Cavalry, about	2,000 „
Artillery, about	2,600 „
Total, about	27,000 men

The Prussians have sent the following contingent:

	Battalions
1.—Combined army corps of Prince Frederick Charles	
Sixth Division:	
11th Brigade, 20th and 60th Regiments	6
12th Brigade, 24th and 64th Regiments	6
Besides the 35th Light Infantry Regiment	3
Thirteenth Division:	
25th Brigade, 13th and 53rd Regiments	6
26th Brigade, 15th and 55th Regiments	6
7th Chasseur Battalion	1
2.—Division of Guards; General Mulbe:	
1st Brigade, 3rd and 4th Foot Guards	6
2nd Brigade, 3rd and 4th Grenadier Guards	6
Chasseurs of the Guard	1
Total	41

Or at 800 men per battalion	32,800 men
Cavalry	3,000 „
Artillery	3,000 „
	38,800 „
With Austrians	27,000 „
Total allied army	65,800 men

Or less than two allied soldiers to one Dane. If the strength of the Danish defences at the Dannevirke, at Düppel, and at Fridericia is taken into account, such a numerical superiority is not more than required to ensure success. It is almost precisely the same proportion of superiority which Wellington and Blücher, in 1815, had over Napoleon.

F. E.

PART VI

THE SEVEN WEEKS' WAR, *1866*

31. Notes on the War, No. 1[1]

THE following notes are intended to comment impartially, and from a strictly military point of view, upon the current events of the war, and, as far as possible, to point out their probable influence upon impending operations.

The locality where the first decisive blows must be struck is the frontier of Saxony and Bohemia. The war in Italy can scarcely lead to any decisive results so long as the Quadrilateral remains untaken, and to take that will be rather a lengthy operation. There may be a good deal of warlike action in Western Germany, but from the strength of the forces engaged, it will be altogether subordinate in its results to the events on the Bohemian frontier. To this neighbourhood, therefore, we shall, for the present, exclusively direct our attention.

In order to judge of the strength of the contending armies it will suffice, for all practical purposes, if we take into account the infantry only, keeping in mind, however, that the strength of the Austrian cavalry will be to the Prussian as three to two. The artillery will be, in both armies, in about the same proportion as the infantry, say three guns per 1,000 men.

The Prussian infantry consists of 253 battalions of the line, 83½ depot battalions, and 116 battalions of the Landwehr (first levy, containing the men from 27 to 32 years of age). Of these, the depot battalions and Landwehr form the garrisons of the fortresses, and are intended, besides, to act against the smaller German states, while the line is massed in and around Saxony to oppose the Austrian army of the north. Deducting about 15 battalions occupying Schleswig-Holstein, and another 15 battalions—the late garrisons of Rastatt, Mainz, and Frankfurt, now concentrated at Wetzlar—there remain about 220 battalions for the

[1] Unsigned article in the *Manchester Guardian*, June 20, 1866.

main army. With cavalry and artillery, and such Landwehr as may be drawn from the neighbouring fortresses, this army will contain about 300,000 men, in nine army corps.

The Austrian army of the north counts seven army corps, each of which is considerably stronger than a Prussian one. We know very little at present of their composition and organization, but there is every reason to believe that they form an army of from 320,000 to 350,000 men. Numerical superiority, therefore, seems assured to the Austrians.

The Prussian army will be under the command-in-chief of the King,—that is to say, of a parade soldier of at best very mediocre capacities, and of weak, but often obstinate, character. He will be surrounded, firstly, by the general staff of the army, under General Moltke, an excellent officer; secondly, by his 'private military cabinet', composed of personal favourites; and thirdly, by such other unattached general officers as he may call to his suite. It is impossible to invent a more efficient system for ensuring defeat at the very headquarters of an army. Here is, at the very beginning, the natural jealousy between the staff of the army and the Cabinet of the King, each of which sections will struggle for supreme influence and will concoct and advocate its own pet plan of operations. This alone would render almost impossible all singleness of purpose, all consistent action. But then come the interminable councils of war, which, in nine cases out of ten, end in the adoption of some half measure—the very worst course in war. The orders of today, in such cases, generally contradict those of yesterday, and when matters become complicated or threaten to go wrong, no orders at all are given out, and things take their own course. '*Ordre, contre-ordre, désordre*,' as Napoleon used to say. Nobody is responsible, because the irresponsible King takes all responsibility upon himself, and, therefore, nobody does anything until distinctly ordered to do so. The campaign of 1806 was commanded in a similar way by the father of the present King; the defeats of Jena and Auerstädt, and the destruction of the whole Prussian army within three weeks, was the consequence. There is no reason to suppose that the present King is superior in mettle to his father; and if he has found in Count Bismarck a man whose political direction he can implicitly follow, there is no man of sufficient standing in the army to take exclusive charge, in a similar way, of military matters.

The Austrian army is under the unconditional command of General Benedek, who is an experienced officer and who, at least, knows his mind. The superiority of supreme command is decidedly on the side of the Austrians.

The Prussian troops are subdivided into two 'armies'; the first, under Prince Frederick Charles, composed of the 1st, 2nd, 3rd, 4th, 7th, and 8th corps; the second, under the Crown Prince, of the 5th and 6th corps. The Guards, forming the general reserve, will probably join the first army. Now this sub-division not only breaks the unity of command, but it also induces, very often, the two armies to move on two different lines of operation, to make combined movements, to lay their mutual point of junction within the reach of the enemy; and in other words, it tends to keep them separated whereas they ought, as much as possible, to keep them together. The Prussians in 1806, and the Austrians in 1859, under very similar circumstances, followed the same course, and were beaten. As to the two commanders, the Crown Prince is an unknown magnitude as a soldier; and Prince Frederick Charles certainly did not show himself to be a great commander in the Danish war.

The Austrian army has no such subdivision; the commanders of the army corps are placed directly under General Benedek. They are, therefore, again superior to their opponents as far as the organization of the army goes.

The Prussian soldiers, especially the men of the reserve and such Landwehr men as had to be taken to fill up vacancies in the line (and there are many) go to war, against their will; the Austrians, on the contrary, have long wished for a war with Prussia, and await with impatience the order to move. They have, therefore, also the advantage in the *morale* of the troops.

Prussian has had no great war for fifty years; her army is, on the whole, a peace army, with the pedantry and martinetism inherent to all peace armies. No doubt a great deal has been done latterly, especially since 1859, to get rid of this; but the habits of forty years are not so easily eradicated, and a great number of incapable and pedantic men must still be found, particularly in the most important places—those of the field officers. Now the Austrians have been fundamentally cured of this complaint by the war of 1859, and have turned their dearly-bought experience to the very best use. No doubt, in organization of detail, in adaptation, for, and experience in, warfare, the Austrians again are superior to the Prussians.

With the exception of the Russians the Prussians are the only troops whose normal formation for fighting is the deep close column. Imagine the eight companies of an English battalion in a quarter-distance column, but two companies instead of one forming the front, so that four rows of two companies each form the column, and you have the 'Prussian column of attack'. A better target for rifled fire-firms than this could not be imagined, and, since rifled cannon can throw a shell into it at 2,000 yards range, such a formation must render it almost impossible to reach the enemy at all. Let one single shell explode in the midst of this mass, and see whether that battalion is fit for anything afterwards on that day.

The Austrians have adopted the loose open column of the French, which is scarcely to be called a column; it is more like two or three lines following each other at 20 or 30 yards distance, and is scarcely, if anything more exposed to losses by artillery than a deployed line. The advantage of tactical formation is, again, on the side of the Austrians.

Against all these advantages the Prussians have but two points to set off. Their commissariat is decidedly better, and the troops will therefore be better fed. The Austrian commissariat, like all Austrian administration, is one den of bribery and peculation scarcely better than in Russia. Even now we hear of the troops being badly and irregularly fed in the field and in the fortresses it will be worse still, and the Austrian Administration may happen to be a more dangerous enemy to the fortresses in the Quadrilateral than the Italian artillery.

The second set-off the Prussians have is their superior armament. Although their rifled artillery is decidedly better than that of the Austrians, this will make very little difference in the open field. The range, trajectory, and accuracy of the Prussian and Austrian rifles will be about on a par; but the Prussians have breech-loaders, and can deliver a steady well-aimed fire in the ranks at least four times a minute. The immense superiority of this arm has been proved in the Danish war, and there is no doubt the Austrians will experience it in a far higher degree. If they, as it is said Benedek has instructed them to do, will not lose much time with firing, but go at the enemy at once with the bayonet, they will have enormous losses. In the Danish war, the loss of the Prussians was never more than one fourth, sometimes only one tenth, of that of the Danes; and, as a military correspondent of *The Times* a short time ago very correctly pointed out, the Danes were almost everywhere beaten by a minority of troops actually engaged.

Still, in spite of the needle gun, the odds are against the Prussians; and if they refuse to be beaten in the first great battle by the superior leadership, organization, tactical formation, and *morale* of the Austrians, and last, not least, by their own commanders, then they must certainly be of a different mettle from that of which a peace army of 50 years' standing may be expected to be.

32. Notes on the War No. 2[1]

People begin to grow impatient at the apparent inactivity of the two great armies on the Bohemian frontier. But there are plenty of reasons for this delay. Both the Austrians and the Prussians are perfectly aware of the importance of the impending collision, which may decide the result of the whole campaign. Both are hurrying up to the front whatever men they can lay their hands on; the Austrians from their new formations (the fourth and fifth battalions of the Infantry regiments), the Prussians from the Landwehr, which at first was intended for garrison duty only.

At the same time, there appears to be on either side an attempt to out-manœuvre the opposing army, and to enter upon the campaign under the most favourable strategical conditions. To understand this, we shall have to look at the map and examine the country in which these armies are placed.

Taking it for granted that Berlin and Vienna are the normal points of retreat of the two armies, and that therefore the Austrians will aim at the conquest of Berlin and the Prussians at that of Vienna, there are three routes by which they might operate. A large army requires a certain extent of country from the resources of which it has to live on the march, and is compelled, in order to move quick, to march in several columns on as many parallel roads; its front will, therefore, be extended on a line which may vary between, say, sixty and sixteen miles, according to the proximity of the enemy and the distance of the roads from each other. This will have to be kept in mind.

[1] Unsigned article in the *Manchester Guardian*, June 25, 1866.

The first route would be on the left bank of the Elbe and Moldau, by Leipzig and Prague. It is evident that on this route each of the belligerents would have to cross the river twice, the second time in the face of the enemy. Supposing either army to attempt to turn, by this route, the flank of its opponent, the latter, having the shorter, because straighter, road, could still anticipate the turning force on the line of the river, and if successful in repelling it, could march straight upon the enemy's capital. This route, equally disadvantageous to both parties, may therefore be dismissed from consideration.

The second route is on the right bank of the Elbe, between it and the Sudeten mountain chain which divides Silesia from Bohemia and Moravia. This is almost on the straight line from Berlin to Vienna; the portion now lying between the two armies is marked out by the railway from Lobau to Pardubitz. This railway passes through that portion of Bohemia which is bounded by the Elbe to the south and west, and the mountains to the north-east. It has plenty of good roads, and if the two armies were to march straight at each other, here would be the point of collision.

The third route is that by Breslau, and thence across the Sudeten chain. This chain, of no considerable elevation, on the Moravian frontier, where it is crossed by several good roads, rises to greater elevation and abruptness in the Riesengebirge, which forms the boundary of Bohemia. Here there are but few roads across; in fact, between Trautenau and Reichenberg, a distance of forty miles, the whole north-eastern portion of the range is not traversed by a single military road. The only road in existence there, that from Hirschberg to the valley of the Iser, stops short at the Austrian frontier. It follows, then, that this whole barrier of forty miles in length, is impassable, at least for a large army, with its innumerable impediments, and that an advance upon or by Breslau must pass the mountains to the south-west of the Riesengebirge.

Now, what are the relative positions of the two armies, with regard to their communications, if engaged on this route?

The Prussians, by advancing due south from Breslau, lay open their communications with Berlin. The Austrians might, if strong enough to command the almost absolute certainty of victory, leave them to advance as far as the intrenched camp of Olmütz, which would stop them, while they themselves could march upon Berlin, trusting to re-open any temporarily-interrupted communications by a decisive victory; or they might meet the Prussian columns singly as they debouch from the mountains, and, if successful, drive them back upon Glogau and Posen, whereby Berlin and the greater portion of the Prussian states would be at their mercy. Thus an advance by Breslau would be advisable for the Prussians in case of a great numerical superiority only.

The Austrians are in a far different position. They have the advantage that the bulk of the monarchy lies south-east of Breslau; that is, in the *direct prolongation* of a line drawn from Berlin to Breslau. Having fortified the northern bank of the Danube near Vienna, so as to shelter the capital from a surprise, they may, temporarily and even for a length of time, sacrifice their direct communication with Vienna, and draw their supplies of men and stores from Hungary. They can, therefore, with equal safety operate by way of Lobau and by way of Breslau, to the north or to the south of the hills; they have far greater freedom in manœuvring than their opponents.

The Prussians, moreover, have further reasons to be cautious. From the northern frontier of Bohemia, the distance to Berlin is not much more than half of that to Vienna; Berlin is so much more exposed. Vienna is sheltered by the Danube, behind which a beaten army can find protection; by the fortifications erected to the north of that river; and by the intrenched camp of Olmütz, which the Prussians could not pass unnoticed with impunity, if the mass of the Austrian army, after a defeat, were to take up a position there. Berlin has no protection of any kind, except the army in the field. Under these circumstances, and those detailed in our first number, the part destined for the Prussians appears to be clearly marked out as a defensive one.

The same series of circumstances, and strong political necessity besides, almost compels Austria to act on the offensive. A single victory may ensure to her great results, while her defeat would not break her power of resistance.

The strategical plan of the campaign in its fundamental features is necessarily very simple. Whichever of the two attacks first, he has only this alternative: either a false attack *north-west* of the Riesengebirge, and the true attack south-east of it, or *vice versa*. The forty-mile barrier is the decisive feature of the seat of war, and round it the armies must gravitate. We shall hear of fighting at both its extremities, and a very few days afterwards will clear up the direction of the true attack, and probably the fate of the first campaign. Yet, with two such unwieldy armies opposed to each other, we feel inclined to think that the most direct route is the safest, and that the difficulty and danger of moving such large bodies of troops in separate columns on different roads through a difficult mountain country, will almost naturally draw both opposing armies on the route Lobau-Pardubitz.

The actual movements which have taken place are as follow: The Prussians, in the first week of June, massed their army of Saxony along the Saxon frontier, from Seitz to Görlitz and their Silesian army from Hirschberg to Neisse. By the 10th June they drew nearer together, having their right wing on the Elbe near Torgau, and their extreme left near Waldenburg. From the 12th to the 16th, the army of Silesia, now consisting of the 1st, 5th, and 6th corps and the Guards, were again extended to the east, this time as far as Ratibor, that is to say, into the extreme south-eastern corner of Silesia. This looks like a feint, especially the parading of the Guards, which are supposed to be always with the main army. If it be more than a feint, or if measures have not been taken to move these four corps back towards Görlitz at the shortest notice and in the shortest time, then this massing of more than 120,000 men in a remote corner is a palpable mistake; they may be cut off from all possibility of retreat and certainly from all connection with the remainder of the army.

Of the Austrians we know little more than that they were concentrated around Olmütz. *The Times* correspondent in their camp states that their sixth corps, 40,000 strong, arrived on the 19th from Weisskirchen at Olmütz indicating a movement to the westward. He adds that on the 21st headquarters were to be shifted to Trübau, on the frontier between Moravia and Bohemia. This move would point in the same direction, if it did not look exceedingly like a *canard* sent on to London with the intention of being thence telegraphed to the Prussian headquarters in order to mislead them. A general who acts with such secrecy as Benedek, and who has such objections to newspaper correspondents, is not likely to inform them on the 19th where his headquarters will be on the 21st, unless he has his reasons for it.

Before concluding, we may be allowed to cast a glance at the operations in North-western Germany. The Prussians had more troops here than was at first known. They had 15 battalions disposable in Holstein, 12 in Minden, and 18 in Wetzlar. By rapid concentric moves, during which the troops showed a quite unexpected capability of supporting forced marches, they took possession in two days of all the country north of a line from Coblenz to Eisenach, and of every line of communication between the eastern and western provinces of the kingdom. The Hessian troops, about 7,000 strong, managed to escape, but the Hanoverians, 10,000 or 12,000, had their direct line of retreat towards Frankfurt cut off, and already on the 17th the rest of the 7th Prussian army corps, 12 battalions, together with the two Coburg battalions arrived in Eisenach from the Elbe. Thus the Hanoverians appear to be hemmed in on all sides, and could escape only by a miracle of stupidity on the part of the Prussians. As soon as their fate will be settled, a force of 50 Prussian battalions will be available against the Federal army which Prince Alexander of [Hesse-]Darmstadt is forming at Frankfurt, and which will consist of about 23,000 Württembergers, 10,000 Darmstaders, 6,000 Nassauers, 13,000 Badeners (only mobilizing now), 7,000 Hessians, and 12,000 Austrians, now on the road from Salzburg; in all about 65,000 men, who may be possibly reinforced by from 10,000 to 20,000 Bavarians. About 60,000 men of these are now reported as already concentrated at Frankfurt, and Prince Alexander has ventured upon a forward move by re-occupying Giessen on the 22nd. This, however, is of no consequence. The Prussians will not advance against him until they are well concentrated, and then, with 70,000 men of all arms, and their superior armament, they ought to make short work of this motley army.

33. Notes on the War, No. III[1]

The first great battle has been fought, not in Bohemia, but in Italy, and the Quadrilateral has again given the Italians a lesson in strategy. The strength of this famous position, as indeed of all fortified positions of any value, consists, not so much in the high defensive capabilities of its four fortresses, but in their being so grouped in a country with strongly-marked military features that the attacking force is almost always induced, and often compelled, to divide itself and attack on two different points,

[1] Unsigned article in the *Manchester Guardian*, June 28, 1866.

while the defending force can send its whole combined strength against one of these attacks, crush it by superior numbers, and then turn against the other. The Italian army has been induced to commit this fault. The King stood with eleven divisions on the Mincio, while Cialdini with five divisions faced the Lower Po, near Ponte Lagoscuro and Polesella. An Italian division counts 17 battalions of 700 men each; consequently, Victor Emmanuel would have, with cavalry and artillery at least 120,000 or 125,000 men, and Cialdini about half that number. While the King crossed the Mincio on the 23rd, Cialdini was to cross the Lower Po and act upon the rear of the Austrians; but up to the moment we write, no certain news have arrived of this latter movement having been effected. At all events, the 60,000 men whose presence might, and probably would, have turned the scale on Sunday last at Custozza, cannot so far have obtained any advantage at all commensurate to the loss of a great battle.

The Lake of Garda lies encased between two spurs of the Alps, forming, to the south of it, two clusters of hills, between which the Mincio forces its way towards the lagoons of Mantua. Both of these groups form strong military positions; their slopes towards the south overlook the Lombard plain, and command it within gun-range. They are well known in military history. The western group, between Peschiera and Lonato, was the scene of the battles of Castiglione and Lonato in 1797, and of Solferino in 1859; the eastern group, between Peschiera and Verona, was contested during three days in 1848, and again in the battle of last Sunday.

This eastern group of hills slopes down on one side towards the Mincio, where it ends in the plain at Valleggio; on the other side, in a long arc, facing south-east, towards the Adige, which it reaches at Bussolengo. It is divided, from north to south, in two about equal portions by a deep ravine, through which flows the rivulet Tione; so that a force advancing from the Mincio will have first to force the passage of the river, and immediately afterwards find itself again arrested by this ravine. On the edge of the slope, facing the plain, and east of the ravine, are the following villages: Custozza, on the southern extremity; further north, in succession, Somma Campagna, Sona, and Santa Giustina. The railway from Peschiera to Verona crosses the hills at Somma Campagna, the high road at Sona.

In 1848, after the Piedmontese had taken Peschiera, they blockaded Mantua and extended their army from beyond that place to Rivoli, on the Lake of Garda, their centre occupying the hills in question. On the 23rd July Radetzky advanced with seven brigades from Verona, broke through the centre of this over-extended line, and occupied the hills himself. On the 24th and 25th the Piedmontese tried to re-take the position, but were decisively beaten on the 25th, and retreated at once through Milan beyond the Ticino. This first battle of Custozza decided the campaign of 1848.

The telegrams from the Italian headquarters about last Sunday's battle are rather contradictory; but, with the assistance of those from the other side, we get a pretty clear insight into the circumstances under which it was fought. Victor Emmanuel intended his 1st corps (General Durando, four divisions or 68 battalions), to take up a position between Peschiera and Verona, so as to be able to cover a siege of the former place. This position must, of course, be Sona and Somma Campagna. The 2nd corps (General Cucchiari, three divisions or 51 battalions) and 3rd corps (General Della Rocca, of the same strength as the second) were to cross the Mincio at the same time, to cover the operations of the 1st. The 1st corps must have crossed near or south of Saliongo, and taken the road of the hills at once, the 2nd seems to have crossed at Valleggio, and the 3rd at Goito, and advanced in the plain. This took place on Saturday the 23rd. The Austrian brigade Pulz, which held the outposts on the Mincio, fell slowly back on Verona; and on Sunday, the anniversary of Solferino, the whole of the Austrian army debouched from Verona to meet the enemy. They appear to have arrived in time to occupy the heights of Sona and Somma Campagna, and the eastern edge of the ravine of the Tione before the Italians. The struggle then would principally be fought for the passage of the ravine. At the southern extremity the two corps in the plain could co-operate with the 1st Italian corps in the hills, and thus Custozza fell into their hands. Gradually the Italians in the plain would advance more and more in the direction of Verona, in order to act upon the Austrian flank and rear, and the Austrians would send troops to meet them. Thus the front lines of the two armies which were originally facing east and west respectively, would wheel round a quarter circle, the Austrians facing south and the Italians north. But, as the hills retreat from Custozza to the north-east, this flank movement of the Italian 2nd and 3rd corps could not immediately affect the position of their 1st corps in the hills, because it could not be extended far enough without danger to the flanking troops themselves. Thus the Austrians appear merely to have occupied the 2nd and 3rd corps by troops sufficient to break their first impetus, while they launched every available man upon the 1st corps, and crushed it by superior numbers. They were perfectly successful; the first corps was repulsed, after a gallant struggle, and at last Custozza was stormed by the Austrians. By this, the Italian right wing advanced east and north-east beyond Custozza, appears to have been seriously endangered; consequently a new struggle for the village took place, during which the lost connection must have been restored, and the Austrian advance from Custozza checked, but the place remained in their hands, and the Italians had to re-cross the Mincio the same night.

We give this sketch of the battle, not as a historical account—for which every detail is as yet wanting—but merely as an attempt, map in hand, to reconcile the various telegrams relating to it amongst each other, and with military common sense; and if the telegrams were anything like correct and complete, we feel confident that the general outline of the battle would appear to be not very different from what we have stated.

The Austrians lost from 600 prisoners, the Italians 2,000, and a few guns. This shows the battle to have been a defeat, but no disaster. The forces must have been pretty equally matched, although it is very probable that the Austrians had less troops under fire than their opponents. The Italians have every reason to congratulate themselves that they were not driven back into the Mincio; the position of the 1st corps between that river and the ravine, on a strip of land between two and four miles wide, and a superior enemy in front, must have been one of considerable danger. It was undoubtedly a mistake to send the main body of the troops into the plain; while the commanding heights, the decisive points, were neglected; but the greatest mistake was, as we pointed out before, to divide the army, to leave Cialdini with 60,000 men on the Lower Po, and to attack with the remainder alone. Cialdini could have contributed to a victory before Verona, and then, marching back to the Lower Po, have effected his passage much more easily, if this combined manœuvre was to be insisted upon at all hazards. As it is, he seems no further advanced than on the first day, and may now have to meet stronger forces than hitherto. The Italians ought, by this time, to know that they have a very tough opponent to deal with. At Solferino, Benedek, with 26,000 Austrians, held the whole Piedmontese army of fully double that number at bay for the whole day, until he was ordered to retreat in consequence of the defeat of the other corps by the French. That Piedmontese army was much superior to the present Italian army; it was better schooled, more homogeneous, and better officered. The present army is but of very recent formation and must suffer from all the disadvantages inherent to such; while the Austrian army of to-day is much superior to that of 1859. National enthusiasm is a capital thing to work upon, but until disciplined and organized, nobody can win battles with it. Even Garibaldi's 'Thousand' were not a crowd of mere enthusiasts, they were drilled men who had learnt to obey orders and to face powder and shot in 1859. It is to be hoped that the staff of the Italian army, for their own good, will refrain from taking liberties with an army which—if numerically inferior—is intrinsically superior to theirs, and, moreover, holds one of the strongest positions in Europe.

34. Notes on the War, No. IV[1]

Suppose a young Prussian ensign or cornet, under examination for a lieutenancy, to be asked what would be the safest plan for a Prussian army to invade Bohemia? Suppose our young officer were to answer,—'Your best way will be to divide your troops into two about equal bodies, to send one round by the east of the Riesengebirge, the other to the west, and effect their junction in Gitschin.' What would the examining officer say to this? He would inform the young gentleman that this plan sinned against the two very first laws of strategy: Firstly, never to divide your troops so that they cannot support each other, but to keep them well together; and, secondly, in case of an advance on different roads, to effect the junction of the different columns at a point which is not within reach of the enemy; that, therefore, the plan proposed was the very worst of all; that it could only be taken into consideration at all in case Bohemia was quite

[1] Unsigned article in the *Manchester Guardian*, July 3, 1866.

unoccupied by hostile troops, and that, consequently, an officer proposing such a plan of campaign was not fit to hold even a lieutenant's commission.

Yet, this is the very plan which the wise and learned staff of the Prussian army have adopted. It is almost incredible; but it is so. The mistake for which the Italians had to suffer at Custozza, has been again committed by the Prussians, and under circumstances which made it ten-fold worse. The Italians knew at least that, with ten divisions, they would be numerically superior to the enemy. The Prussians must have known that if they kept their nine corps together they would be at best barely on a par, as far as numbers went, with Benedek's eight corps; and that by dividing their troops they exposed the two armies to the almost certain fate of being crushed in succession by superior numbers. It would be completely inexplicable how such a plan could ever be discussed, much less adopted, by a body of such unquestionably capable officers as form the Prussian staff—if it was not for the fact of King William being in chief command. But nobody could possibly expect that the fatal consequences of kings and princes taking high command would come out so soon and so strong. The Prussians are now fighting, in Bohemia, a life-and-death struggle. If the junction of the two armies at or about Gitschin is prevented, if each of the two, being beaten, has to retire out of Bohemia, and, by retiring, to get further away again from the other, then the campaign may be said to be virtually over. Then Benedek may leave the army of the Crown Prince unnoticed while it retires towards Breslau, and follow up, with all his forces, the army of Prince Frederick Charles, which can hardly escape utter destruction.

The question is, Will this junction have been prevented? Up to the moment we write we have no news of events later than Friday evening, the 29th. The Prussians, beaten out of Gitschin (the name of the place, in Bohemian, is spelt Jicin) on the 28th by General Edelsheim, claim to have stormed the town again on the 29th, and this is the last information we possess. The junction was not then effected; at least four Austrian and parts of the Saxon army corps had then been engaged against about five or six Prussian corps.

The various columns of the army of the Crown Prince, as they descended into the valley on the Bohemian side of the hills, were met by the Austrians at favourable points where the valley, widening out, allowed them to offer a larger front to the Prussian columns, and to attempt to prevent them from deploying; while the Prussians would send troops, wherever practicable, through the lateral valleys, to take their opponents in flank and rear. This is always the case in mountain warfare, and accounts for the great number of prisoners that are always made under such circumstances. On the other side, the armies of Prince Frederick Charles and Herwarth von Bittenfeld appear to have got through the passes almost unopposed; the first engagements took place on the line of the Iser river, that is almost midway between the starting points of the two armies. It would be idle to try to disentangle and bring into harmony the fearfully contradictory, and often totally unauthenticated, telegrams which have come to hand these last three or four days.

The fighting has been necessarily very much chequered in its results; as new forces came up, victory favoured first one and then the other side. Up to Friday, however, the general result appears to have been, so far, in favour of the Prussians. If they maintained themselves in Gitschin, no doubt the junction was effected on Saturday or Sunday, and then their greatest danger would be passed. The final fight for the junction would probably be fought with concentrated masses on both sides, and decide the campaign for some time, at least. If the Prussians were victorious, they would be at once out of all their self-begotten difficulties, but they might have obtained the same, and even greater, advantages without exposing themselves to such unnecessary dangers.

The fighting appears to have been very severe. The very first Austrian brigade which met the Prussians in battle, was the 'black and yellow' brigade, which, in Schleswig, stormed the Königsberg, near Oberselk, the day before the evacuation of the Dannevirke. It is called black and yellow after the facings of the two regiments composing it, and was always considered one of the best brigades in the service. They were, however, beaten by the needle-gun, and above 500 men of one of its regiments (Martini) were taken prisoners after they had charged the Prussian lines five times in vain. In a later engagement, the colours of the 3rd battalion of the Deutschmeister regiment were taken. This regiment, recruited in Vienna exclusively, is considered the best in the whole army. Thus the very best troops have been already in action. The Prussians must have behaved splendidly for an old peace army. When war was actually declared, a totally different spirit came over the army, brought on, chiefly, by the clearing-out of the small fry of potentates in the north-west. It gave the troops—rightly or wrongly, we merely register the fact—the idea that they were asked to fight, this time, for the unification of Germany, and the hitherto sullen and sulky men of the reserve and Landwehr then crossed the frontier of Austria with loud cheers. It is owing to this chiefly that they fought so well; but at the same time we must ascribe the greater portion of whatever success they have had to their breech-loaders; and if they ever get out of the difficulties into which their generals so wantonly placed them, they will have to thank the needle-gun for it. The reports as to its immense superiority over the muzzle-loaders are again unanimous. A sergeant from the Martini regiment, taken prisoner, said to the correspondent of the *Cologne Gazette*: 'We have surely done whatever may be expected from brave soldiers, but no man can stand against that rapid fire.' If the Austrians are beaten, it will be not so much General Benedek or General Ramming as General Ramrod who is to blame for the result.

In the north-west, the Hanoverians, brought to a sense of their position by a sharp attack from General Manteuffel's advanced guard under General Flies, have surrendered, and thereby 59 Prussian battalions will be at liberty to act against the Federal troops. It was high time, too, that this should be done before Bavaria had completed all her armaments, for otherwise much stronger forces would be required to subdue South-western Germany. Bavaria is notoriously always slow and behindhand with her military arrangements, but when they are complete, she can bring into the field from 60,000 to 80,000 good troops. We may now soon hear of a rapid concentration of Prussians on the Main and of active operations against Prince Alexander of Hesse-Darmstadt and his army.

35. Notes on the War, No V[1]

The campaign which the Prussians opened with a signal strategic blunder has been since carried on by them with such a terrible

[1] Unsigned article in the *Manchester Guardian*, July 6, 1866.

tactical energy that it was brought to a victorious close in exactly eight days.

We said in our last note that the only case in which the Prussian plan of invading Bohemia by two armies separated by the Riesengebirge could be justified was that in which Bohemia was unoccupied by hostile troops. The mysterious plan of General Benedek appears to have mainly consisted in creating a situation of that sort. There appear to have been but two Austrian army corps—the 1st (Clam Gallas) and the 6th (Ramming)—in the north-western corner of Bohemia, where, from the beginning, we expected the decisive actions would be fought. If this was intended to draw the Prussians into a trap, Benedek has succeeded so well that he got caught in it himself. At all events, the Prussian advance on two lines, with from forty to fifty miles of impassable ground between them, towards a point of junction two full marches from the starting points, and within the enemy's lines,—this advance remains a highly dangerous manœuvre under all circumstances, and one which would have been followed by signal defeat but for Benedek's strange slowness, for the unexpected dash of the Prussian troops, and for their breech-loading rifles.

The advance of Prince Frederick Charles took place with three corps (the 3rd, 4th, and 2nd, the latter in reserve), by Reichenberg, north of a difficult range of hills, on the southern side of which General Herwarth advanced with a corps and a half (the 8th and one division of the 7th). At the same time, the Crown Prince stood, with the 1st, 5th, and 6th corps, and the Guards, in the mountains about Glatz. Thus the army was divided into three columns—one on the right, of 45,000, one in the centre, of 90,000, and one on the left, of 120,000 men—none of which could support either of the others for at least several days. Here, if ever, there was a chance for a general commanding at least an equal number of men to crush his opponents in detail. But nothing appears to have been done. In the 26th Prince Frederick Charles had the first serious engagement, at Turnau, with a brigade of the 1st corps, by which he established his communication with Herwarth; on the 27th, the latter took Münchengratz, while, of the army of the Crown Prince, a first column, the 5th corps, advanced beyond Nachod, and beat the 6th Austrian corps (Ramming) severely; on the 28th, the only slightly unlucky day for the Prussians, Prince Frederick Charles's advance guard took Gitschin, but was again dislodged by General Edelsheim's cavalry, while the 1st corps of the army of the Crown Prince was checked with some loss at Trautenau by the 10th Austrian corps of Gablenz, and only disengaged by the advance of the Guards towards Eipel, on an intermediate road between the 1st and 5th Prussian corps. On the 29th, Prince Frederick Charles stormed Gitschin, and the army of the Crown Prince totally defeated the 6th, 8th, and 10th Austrian corps. On the 30th, a fresh attempt of Benedek's to re-take Gitschin by the 1st corps and the Saxon army was signally foiled, and the two Prussian armies effected a junction. The Austrian loss represents men to the number of at least a corps and a half, while that of the Prussians is less than one fourth that number.

Thus we find that on the 27th there were only two Austrian army corps, of about 33,000 men each, at hand; on the 28th, three; on the 29th, four, and if one Prussian telegram be correct, part of a fifth (the 4th corps); while on the 30th the Saxon army corps only had been able to come up in support. There were, then, two, if not three, corps absent from the contested ground during all that time, while the Prussians brought every man down into Bohemia. In fact, up to the evening of the 29th, the whole of the Austrian troops on the spot were barely superior in numbers to either of the two Prussian armies, and being brought into line successively, the supports arriving after the defeat only of the troops first engaged, the result was disastrous.

The 3rd army corps (Archduke Ernest), which fought at Custozza, is reported to have been sent to the north by rail immediately after that battle, and is, in some accounts, set down among Benedek's available forces. This corps, which would make the whole force, including the Saxons, nine corps, could not have been up in time for the battles in the latter days of June.

The Prussians, whatever the faults of their plan of operations were, made up for them by their rapidity and energy of action. No fault can be found with the operations of either of their two armies. Short, sharp, and decisive were all their blows, and completely successful. Nor did this energy forsake them after the two armies were joined; on they marched, and already on the 3rd they met Benedek's combined forces with the whole of theirs, and gave them a last crushing blow.

It is hardly possible to suppose that Benedek accepted this battle of his own free will. No doubt the rapid pursuit of the Prussians compelled him to take a strong position with all his army, in order to re-form his troops, and to give a day's start to his retiring army train, expecting not to be attacked in force during the day, and to be able to draw off during the night. A man in his position, with four of his corps completely shattered, and after such tremendous losses, cannot have desired, there and then, to deliver a decisive battle, if he could draw off in safety. But the Prussians appear to have compelled him to fight, and the result was the complete rout of the Austrians, who, if the armistice be not granted, will now be trying to make towards Olmütz or Vienna, under the most disadvantageous circumstances, for the slightest out-flanking movement of the Prussians on their right must cut off numerous detachments from the direct road, and drive them into the hills of Glatz, to be made prisoners. The 'army of the north', as splendid a host as there was in Europe ten days ago, has ceased to exist.

No doubt the needle-gun, with its rapid fire, has done a great part of this. It may be doubted whether without it the junction of the two Prussian armies could have been effected; and it is quite certain that this immense and rapid success could not have been obtained without such superior fire, for the Austrian army is habitually less subject to panic than most European armies. But there were other circumstances co-operating. We have already mentioned the excellent dispositions and unhesitating action of the two Prussian armies, from the moment they entered Bohemia. We may add that they also deviated, in this campaign, from the column system, and brought their masses forward principally in deployed lines, so as to bring every rifle into activity, and to save their men from the fire of artillery. We must acknowledge that the movements both on the march and before the enemy were carried out with an order and punctuality which no man could have expected from an army and administration covered with the rust of fifty years' peace. And, finally, all the world must have been surprised at the dash displayed by these young troops in each and every engagement without exception. It is all very well to say the breech-loaders did it, but they are not self-acting, they want stout hearts and strong arms to carry them. The Prussians fought very often against superior numbers, and were almost everywhere the attacking party; the Austrians, therefore, had the choice of ground. And in attacking strong positions and barricaded towns, the advantages of the breech loader almost disappear; the bayonet has to do the work, and there has been a good deal of it. The cavalry, moreover, acted with the same dash, and with them cold steel and speed of horses are the only weapons in a charge. The French *canards* of Prussian cavalry lines first peppering their opponents with carbine fire (breech-loading or otherwise) and then rushing at them sword in hand, could only originate among a people whose cavalry has very often been guilty of that trick, and always been punished for it by being borne down by the superior impetus of the charging enemy. There is no mistaking it, the Prussian army has, within a single week, conquered a position as high as ever it held, and may well feel confident now to be able to cope with any opponent. There is no campaign on record where an equally signal success, in an equally short time, and without any noteworthy check, has been obtained, except that campaign of Jena which annihilated the Prussians of that day, and, if we except the defeat of Ligny, the campaign of Waterloo.

INDEX